Self-Assessment for Distribution System Optimization

**American Water Works
Association**

Self-Assessment for Distribution System Optimization

Senior Managing Editor/Project Manager: Melissa Valentine
Cover Art: Melanie Yamamoto
Production: PerfecType | Nashville, TN

Library of Congress Cataloging-in-Publication Data
Names: Martin, Barbara (Barbara Stricos), 1975- author.
Title: Self-assessment for distribution system optimization / by Barb Martin.
Description: Denver, CO : American Water Works Association, [2018] | Includes
    bibliographical references.
Identifiers: LCCN 2017049428 | ISBN 9781625762511
Subjects: LCSH: Water--Distribution--Handbooks, manuals, etc.
Classification: LCC TD481 .M368 2018 | DDC 363.6/10684--dc23 LC record available
at https://lccn.loc.gov/2017049428

ISBN: 978-1-62576-251-1
eISBN: 978-1-61300-438-8

American Water Works Association

6666 West Quincy Avenue
Denver, CO 80235-3098
303.794.7711
www.awwa.org

Partnership for Safe Water Program Information

The Partnership for Safe Water (PSW) is a voluntary utility program for drinking water treatment plant and distribution system optimization. The Partnership's mission is to improve the quality of drinking water delivered to customers by optimizing water system operations. This is achieved by completing a comprehensive self-assessment of treatment plant performance and operations, developing and implementing an action plan for improvement, and continuously monitoring progress toward optimization.

The contents of this guide represent the Partnership for Safe Water's approach to surface water treatment plant self-assessment and optimization. While any plant may complete the self-assessment process by following the steps outlined in this guide, utilities are encouraged to subscribe to the Partnership for Safe Water for access to additional resources, recognition, and other subscriber-only benefits.

The Partnership for Safe Water welcomes water utilities, worldwide, to join the hundreds of Partnership subscribers that have benefitted from the program's unique optimization process for more than 20 years. Access additional information about the Partnership for Safe Water at www.awwa.org/partnership.

TABLE OF CONTENTS

LIST OF TABLES

LIST OF FIGURES

FOREWORD

Self-Assessment for Distribution System Optimization is the Partnership for Safe Water's primary guidance document for the distribution system optimization program. Originally published as a draft document in 2011 to fulfill the requirements of a US Environmental Protection Agency (USEPA) contract, the draft has been revised based on input from Partnership for Safe Water volunteers and the feedback of utilities that have completed the distribution system self-assessment process. Utility feedback was a valuable component of the revision process and helped to create a practical, operations-oriented resource designed to support the distribution system optimization efforts of drinking water utility staff.

This guide follows the Partnership's self-assessment approach, also applied in the self-assessment resources, *Self-Assessment for Water Treatment Plant Optimization* and *Self-Assessment for Wastewater Treatment Plant Optimization*, the respective guidance documents for the Partnership for Safe Water and Partnership for Clean Water's treatment plant optimization programs. The basis for this approach remains the proven USEPA Comprehensive Performance Evaluation framework, which provides a comprehensive and systematic basis for assessing utility operations and performance. Results from Water Research Foundation Report #4109, Criteria for Optimized Distribution Systems (Friedman et al. 2010), provide the framework for this document and serve as a suggested primary reference document for readers seeking additional detail or background information.

The primary author of the original (2011) version of this guidance was William C. Lauer. Partnership for Safe Water Steering Committee representatives and utility optimization experts participated in both its initial development and revisions. The Steering Committee provided approval for use of this resource as the distribution system optimization program's primary self-assessment resource.

The Partnership for Safe Water appreciates the efforts of the authors and reviewers who contributed to the development of this significant program resource, as well as those of the utilities that have successfully completed, and benefitted from, the distribution system self-assessment process.

2016 Revision Team
Cynthia Andrews-Tate—Long Beach Water Department, Long Beach, Calif.
Mary Armacost—Speedway Utilities, Speedway, Ind.
Michael Barsotti—Champlain Water District, South Burlington, Vt.
Todd Brewer, PhD, PE—City Utilities of Springfield, Springfield, Mo.
Edgar Chescattie—Pennsylvania Department of Environmental Protection,
  Harrisburg, Pa.
Alex Gerling—AWWA, Denver, Colo.
Brian Haws, PE—Austin Water Utility, Austin, Tex.
Kevin Linder—Aurora Water, Aurora, Colo.
Barbara Martin—AWWA, Denver, Colo.
Michael Salas—Long Beach Water Department, Long Beach, Calif.
Tom Schippert—AWWA, Denver, Colo.
Michael Sullivan—Oak Creek Water & Sewer Utility, Oak Creek, Wis.
Jack Wang, PhD—California American Water, Monterrey, Calif.
Timothy Wilson—Marshalltown Water Works, Marshalltown, Iowa
Yan Zhang, PhD, PE—Long Beach Water Department, Long Beach, Calif.

2011 Working Draft Authors
Mary Armacost—Speedway Water, Speedway, Ind. (Chapter 7)
Michael Barsotti—Champlain Water District, South Burlington, Vt. (Chapter 3)
David Hartman—Cincinnati, Ohio (Chapter 3)
William C. Lauer—Denver, Colo. (Primary Author)
Kevin Linder—Aurora Water, Aurora, Colo. (Chapter 5)
Tom Ries—Aurora Water, Aurora, Colo. (Chapter 6)
William J. Soucie—Central Lake County Joint Action Water Agency, Ill. (Chapter 2)
Jack Wang, PhD—California American Water, Monterrey, Calif. (Chapter 4)

2011 Working Draft Review Committee
Patty Barron—Birmingham Water Works Board, Birmingham, Ala.
Douglas Beck—Lancaster, Pa.
Monty Bishop—Chattanooga, Tenn.
Jeff Chapman—Greenwood Commissioners of Public Works, Greenwood, S.C.
Hubert Demard—Quebec City, Que.
John Diemert—Lake County Department of Utilities, Madison, Ohio
Joseph DeVito—Beaufort Jasper Water & Sewer Authority, Okatie, S.C.
Robert Donnelly—Bear Creek Township, Pa.
Stephen Estes-Smargiassi—Massachusetts Water Resources Authority, Boston,
  Mass.
James W. Fay—Champlain Water District, South Burlington, Vt.
Melinda Friedman—Confluence Engineering Group, LLC, Seattle, Wash.
Dahlia Ghobrial—Passaic Valley Water Commission, Totowa, N.J.

2011 Working Draft Review Committee (continued)

Michael Grimm—West Slope Water District, Portland, Ore.

Paul Handke—Pennsylvania Department of Environmental Protection, Harrisburg, Pa.

Steve Hubbs—Louisville, Ky.

Ronald Hunsinger—Oakland, Calif.

Oleg Kostin—New Jersey American Water, Bound Brook, N.J.

George Kunkel—Philadelphia, Pa.

Kevin Linder—Aurora Water, Aurora, Colo.

Mary Ann Mann—Murietta, Calif.

David G. Miller—Manchester Water Works, Manchester, N.H.

Richard Mosier—Brodhead Creek Regional Authority, East Stroudsburg, Pa.

Jay Nadeau—Champlain Water District, South Burlington, Vt.

Kanwal J. Oberoi—Charleston Water System, Charleston, S.C.

Elizabeth Owens—Wichita Water Utilities, Wichita, Kan.

David C. Peters—Huntington, W.Va.

Steve Price—City of Goose Creek, Goose Creek, S.C.

Marsha Pryor—Pinellas County Utilities, Largo, Fla.

Keith Ristinen—Reno, Nev.

Mark Romain—West Milford, N.J.

Thomas Rothermich—St. Louis, Mo.

Jan Routt—Jan Routt & Associates, LLC, Lexington, Ky.

Chet Shastri—Fort Wayne City Utilities, Fort Wayne, Ind.

Michael Smith—Stafford County Department of Utilities, Stafford, Va.

Mark Stanley—Salt Lake City Department of Public Utilities, Salt Lake City, Utah

Mike Sullivan—Oak Creek Water & Sewer Utility, Oak Creek, Wis.

Yong Wang—City of Houston, Houston, Tex.

Partnership for Safe Water Steering Committee (2011-2017)
Michael Barsotti—Chair (Starting 2015)
AWWA Representative
Champlain Water District, Vt.

Robert Cheng, PhD, PE—Chair (2009-2015)
AWWA Representative
Long Beach Water Department, Calif.

Greg Carroll
USEPA Representative (Starting 2011)
Office of Ground Water and Drinking Water
Cincinnati, Ohio

Brian Haws (2010–2016)
AMWA Representative
Austin Water Utility, Tex.

Andrea Song (Starting 2016)
AMWA Representative
Denver Water, Denver, Colo.

Edgar Chescattie (2006-2012)
ASDWA Representative
Pennsylvania Department of Environmental Protection
Harrisburg, Pa.

Doug Kinard (Starting 2012)
ASDWA Representative
South Carolina Department of Health and Environmental Control
Columbia, S.C.

Keith Cartnick (2007-2013)
NAWC Representative
United Water New Jersey, N.J.

Bruce Hauk (2013-2017)
NAWC Representative
Illinois American Water, Ill.

EXECUTIVE SUMMARY

In the time since the original version of the Partnership for Safe Water's distribution system self-assessment guidance was first published in 2011, the Partnership's distribution system optimization program has doubled in size, and more than 20 utilities have successfully completed the distribution system self-assessment process.

Because the self-assessment process is intended to be adaptable and utility-specific, the utilities that have completed the process have benefitted from it in a variety of ways. Some of these include the improved teamwork, cohesiveness, and communications resulting from utility staff working toward a common goal. Others have benefitted from identifying and centralizing key distribution system performance data as a result of the self-assessment process. For some utilities, the distribution system self-assessment process has proven to be a career development opportunity for young professionals, exposing them to all aspects of distribution system operation as well as the highest levels of utility management. However, staff from all utilities that have completed the distribution system self-assessment process agree that it represents a unique opportunity for system-specific learning and improvement, allowing them to focus on the key operational and performance indicators for distribution system optimization.

Additionally, the utilities that have completed the self-assessment process represent a variety of distribution system sizes and configurations. The self-assessment process has been successfully completed at systems using free chlorine as a residual disinfectant, chloramine as a residual disinfectant, as well as mixed disinfectants. The process has been applied successfully by retail systems, water wholesalers, and consecutive systems serving utility populations ranging from less than 15,000 to well over one million. The combined total miles of pipe represented by these utilities, if laid end to end, would more than circle the earth.

These are truly impressive statistics for a relatively young program—and if they were achieved using the original version of the guidance, why was there a need to update an already effective work?

The answer lies in the current state of knowledge, science, and research pertaining to drinking water distribution systems. With so many recent advances with the potential to improve distribution system assessment, operation, and performance, it is part of the mission of the Partnership for Safe Water to provide this knowledge to its utility subscribers. It is the Partnership's intention that utilities apply the knowledge and procedures presented in this guide to support utility-specific learning and realize distribution system improvements.

Barbara Martin
Director—Partnership Programs
AWWA
Denver, Colo.

INTRODUCTION

Why Optimize?

Drinking water suppliers strive to provide the highest quality in their delivered product. The supplier's job is to proactively protect the public health of the customers it serves. Suppliers must be very tenacious in their approach to this goal, but how can drinking water suppliers really document that they are accomplishing this very important task?

The simple answer to the question of "Why optimize?" is to join the Partnership for Safe Water (PSW) program. Using the program's very practical tools allows a water utility to document its success stories to improve water quality while improving customer confidence in the supplied drinking water. The only way to improve water quality is to fully understand the utility's current conditions. The mission of the PSW program is to guide a water supplier through this self-examination process with tools that the program has optimized over more than 20 years. Since the program's inception in 1995, a population of more than 100 million in North America is being supplied safer drinking water because of utility participation in the PSW.

Numerous award-winning Partnership utilities can attest that the program's "For Utilities by Utilities" approach is an extremely practical way to improve water quality in a very user-friendly, nonregulatory manner. For utility staff, this approach quickly becomes second nature and inherent in day-to-day operations. The program's annual data review process allows a utility to monitor progress while striving to continually improve. The incremental changes and improvements made in process control actually become infectious, and therefore easily transferrable to the entire organization.

The foundation of this voluntary improvement program is the self-assessment process. Through this process, a utility evaluates its present strengths and weaknesses. Each utility can tailor this process to fit its own staff resources and experience based on a schedule set by the utility. The key is to take small incremental steps to make improvements using the very practical program tools. The PSW data collection software allows a utility to collect data and analyze operations very easily, thus allowing fluctuations in day-to-day operations to be reviewed and improved. Over time, this approach allows the participating utilities to see trends and develop prioritized action plans that they manage and schedule. This self-assessment process is the key to the program's success over the past 20 years, and it is proven to work for utilities of any size.

This self-evaluation and assessment process is the heart of the PSW program. This program also has a very talented group of volunteers that reviews the submitted Self-Assessment Completion Reports. With this approach, the people who are most familiar with day-to-day utility operations assist participating utilities throughout the entire process. This makes the goal of continuous quality improvement much easier and attainable for utilities of any size. The PSW program objective is to make small incremental improvements in day-to-day operations toward achieving continuous quality improvements protective of public health. Why not join the best program available to accomplish this critical task?

Partnership for Safe Water Background

The PSW is a voluntary continuous improvement program that uses optimization methods to improve drinking water utility operation and performance. The PSW's mission is to help utilities improve the quality of drinking water delivered to customers by optimizing water system operation.

Six major drinking water organizations support the PSW in its efforts to carry out its critical mission, including the American Water Works Association, USEPA, the Association of Metropolitan Water Agencies, the Association of State Drinking Water Administrators, the National Association of Water Companies, and the Water Research Foundation. The program is governed by a Steering Committee that consists of a representative from each of the program's partner organizations.

The fundamental approach by the Partnership is to improve performance by optimizing system operations, rather than relying solely on significant capital improvements. This approach has been successfully employed for water treatment plants since the program was introduced in 1995 (Renner & Hegg 1997, USEPA 1998b). This self-assessment guide is an important program resource to assist water utility staff in applying similar principles to the process of assessing and optimizing distribution system operations.

The PSW was originally developed to help well-run utilities treating surface water sources optimize their treatment processes, beyond the requirements of the 1989 Surface Water Treatment Rule (SWTR), to provide added protection against the breakthrough of *Cryptosporidium* oocysts. This program was launched after the 1993 Milwaukee, Wisconsin, outbreak but before development of the USEPA's Interim Enhanced Surface Water Treatment Rule and the later Long-Term 1 & 2 Enhanced Surface Water Treatment Rule (USEPA 1988a, 1989a, 1998a, 2002a). The current PSW treatment plant optimization program has more than

250 utility members representing more than 450 surface water treatment facilities that serve a combined population of more than 85 million people across North America.

The distribution system optimization program parallels the PSW's treatment plant optimization program and self-assessment procedures, which were established during the 1990s (Renner & Hegg 1997). Similar to the optimization program for treatment plant operation, the distribution system optimization program includes four phases:

- Phase I—Commitment
- Phase II—Baseline/annual data reporting
- Phase III—Self-assessment and completion of substantial optimization efforts.
- Phase IV—Demonstrated optimization

Distribution systems successfully completing the self-assessment, including successful peer review of the self-assessment completion report, receive the Directors Award of recognition and benefit from the development of system specific action plans to further improve performance.

Systems that have demonstrated convincing and significant improvements, after achieving the Phase III Directors Award, will receive the highest levels of program recognition including the Presidents Award and Phase IV Excellence Award for Distribution System Operation. These awards are presented to utilities that achieve specific distribution system performance targets and are able to demonstrate optimized distribution system operations. These are challenging goals, but all systems that embrace the process and approach it with tenacity can make progress toward optimization and realize the benefits of improved water quality, operations, and system reliability.

Although awards provide utilities with well-deserved recognition, true success in the Partnership program is realized when staff at all levels of management, operations, and maintenance work as a cohesive team to consistently produce the highest quality water possible. Operational excellence is a mindset, a culture, a process, and a demonstrated tenacity to pursue a goal (optimization) that ultimately has no finish line.

Partnership for Safe Water: A Complementary Program

The PSW is one of several formalized water sector utility continuous improvement programs. Although a complete discussion of all water utility continuous improvement programs is beyond the scope of this work, two of these programs include the AWWA Benchmarking Survey, which compiles utility performance data on an annual basis, and Effective Utility Management, which is a collaborative program for assessing and improving utility management. PSW self-assessment and optimization efforts are intended to be collaborative with a utility's participation in other utility improvement programs, and the results of the self-assessment may be used to support improvement in other areas. A more complete discussion of Effective Utility Management is included in Chapter 6 (Administration), along with descriptions of several key performance indicators included in AWWA's Benchmarking Surveys.

PSW Optimization Philosophy

The PSW helps water treatment plants and distribution systems achieve continuous improvement by optimizing operational performance. Optimization, as defined by the Partnership, means that all of the system processes are being performed at the highest level (all the optimization goals of the Partnership are being continuously achieved).

Self-assessment is a tool used to determine utility optimization status and to identify areas for improvement over time. Systems should engage in the self-assessment process to identify opportunities for improvement and to advance their optimization status. Participating utilities embrace the opportunity to discover areas for improvement and develop and implement an action plan to continuously improve and work toward achieving optimized performance. This action plan is a primary outcome of the self-assessment process.

The program's optimization goals are reflective of the performance and operation of an optimized distribution system. It can require tenacity on the part of utility staff to achieve these challenging goals. Optimization efforts are used to advance system operation and work toward achieving the Partnership's optimization goals.

Continuous improvement is a process. Partnership utilities tirelessly strive for higher levels of performance. Even systems that consistently meet the current optimization goals are encouraged to continue to tenaciously search for ways to improve. The PSW philosophy embraces the quest for excellence in water utility operation to improve the quality of water and reliability of service provided to all users.

The Partnership has developed this optimization program for drinking water distribution system operation based on a fundamental understanding about its applicability. The scope of the program and the limits of its applicability must be defined. Table 1-1 illustrates the working definitions used by the PSW for the terms *distribution system* and *optimized distribution system*. It can be difficult to define these terms exactly without some ambiguity. Specific utility situations may be addressed by Partnership staff and leadership as they arise.

Because of the variety and types of distribution system ownership and responsibility arrangements, it is difficult to strictly define a distribution system for the purposes of clarifying eligibility for participation in the PSW distribution system optimization program. Therefore, for Partnership purposes, a

Table 1-1.   Partnership for Safe Water distribution system definitions

Term	Definition
Distribution system	A system of conveyances that distributes potable water as part of a community water system, including all pipes, storage tanks, pipe laterals, valves, and appurtenances used for delivery.
Optimized distribution system	The system continuously delivers high-quality potable water to all users with ultimate safety and reliability.

utility's distribution system is defined as that for which it has responsibility for the quality of water delivered to users. The distribution system may be operated by a community water system or private corporation. It may be defined as a water wholesaler agency or as a consecutive system. All types of distribution systems can benefit from the completion of the self-assessment and optimization process, and the process should include participation from as many stakeholders as possible. In any of these cases, it is important to identify the true operators and owners of the system. The legally responsible system owner may be a separate entity than the distribution system operator. The individual operator of a distribution system includes personnel with hands-on valve turning/maintenance responsibilities, as well as those with remote monitoring and system operation responsibilities through the supervisory control and data acquisition system. Beyond these definitions, distribution systems eligible for Partnership participation continuously use a disinfectant and provide a disinfectant residual throughout the distribution system.

Potable water distribution systems are commonly used to deliver water to customers for uses other than providing drinking water of the highest quality. Most distribution systems are also used for fire protection and to provide water for industrial uses, which can present objectives that are sometimes in conflict with the ultimate goal of delivering drinking water of the highest quality. The PSW distribution system optimization program focuses solely on the delivery of safe, high-quality potable water. Optimization of the distribution system, using the Partnership approach, is designed to help water utilities achieve this objective. There are also nonpotable water distribution systems. Water reuse or nonpotable distribution systems are encouraged to consider participating in the Partnership for Clean Water program, which provides optimization opportunities for nonpotable applications, such as reuse and wastewater treatment.

The potable water distribution system, as defined by the PSW, is limited to those facilities and components that are owned (or controlled) and operated by the community water system. Utilities may find that collaboration with other entities, for example the local fire department, is necessary to achieve the utility's optimization goals. The need for such collaboration may be evaluated as part of the self-assessment process. It is recommended that the process include input from as many systems stakeholders as possible to obtain the most accurate information about the system's optimization status.

The PSW distribution system optimization program objective is to help utilities consistently deliver high-quality potable water to all users. To accomplish this, a utility must achieve and demonstrate commitment to optimization from staff at all levels of management, operations, and maintenance. Only when staff at *all* levels work as a cohesive and committed optimization team can the utility effectively improve operations and water quality, meet the needs of customers, and ensure compliance with current and future regulations.

Development of This Guide

This guide, and the PSW Distribution System Optimization Program, is the result of more than 10 years of discussions, planning, and research. Hundreds of utility representatives, local and federal regulators, consultants, and other subject matter experts have provided their advice and support to develop a voluntary continuous improvement program for distribution system operation. This guide represents the results from a phased, collaborative process, based on the results from many studies and years of practical experience.

There are many references listed throughout this guide. However, the primary basis for the work is Water Research Foundation report for project #4109 *Criteria for Optimized Distribution Systems* (Friedman et al. 2010). This was a tailored collaboration project funded by the Water Research Foundation, with contribution by the PSW. Much of the content of this work came directly from the project's final report. The self-assessment methods, criteria, and approach developed by this project included input from more than 200 utility representatives and a project team of nearly 150. This broad participation was supplemented by many PSW utilities that reviewed and commented on the content. Further, the contents of this work were approved by the PSW Steering Committee, consisting of a representative from each of the program's six partner organizations.

The literature review included in the project #4109 report (used as a basis for this work) consisted of a search of key distribution system optimization-related research reports, standards, manuals, white papers, regulations, and existing optimization programs and tools. A list of the materials reviewed is provided in Appendix A of the Water Research Foundation report #4109 (Friedman et al. 2010). Besides this extensive literature review, the PSW distribution system self-assessment guide development team included other references, throughout this guide, that support specific statements or conclusions. The list of references included in this guide are one of its many benefits because it provides readers with a compilation of some of the most recent and relevant distribution system optimization resources.

It is important for utilities that are considering participation in this voluntary optimization program to understand that it is meant to be comprehensive and challenging. Similar to the existing PSW program for assessing and optimizing water treatment plants, the numeric criteria established are intended to define excellence rather than reflect existing regulatory requirements, other industry benchmarks, or reported average conditions. It is expected, therefore, that all utilities, even those with high-performing distribution systems, will identify opportunities for improvement and implement system-specific action plans to make targeted improvements to attain the program's optimization goals. Consequently, it is anticipated that using the numeric goals, self-assessment process, software tools, and implementation of the utility-defined action plan will promote continuous and significant utility improvement in distribution system operations and performance.

Regulatory Framework

The need for water utility optimization, from source to tap, and the inherent complexity associated with such an endeavor, has long been recognized. The drinking water sector has made notable progress on the regulation and optimization of surface water supplies. Many regulations are in place that address source and finished water quality, and the Partnership has successfully developed and implemented a self-assessment and optimization program for water treatment plants (Renner & Hegg 1997, USEPA 1998b, Linder & Martin 2015). Although there are several USEPA Safe Drinking Water Act regulations, as well as international regulations, that target distribution system-related parameters (such as total coliform, lead and copper, disinfectants, and disinfection by-products), the various regulatory compliance strategies can result in competing and sometimes conflicting priorities. This is often referred to as the challenge of "simultaneous compliance."

The USEPA, local regulatory agencies, water professionals, and researchers have been involved in substantial work toward the development of distribution system optimization and assessment programs. The focus of this continuing work has been to help utilities evaluate their own distribution system performance and optimization status. This is achieved by benchmarking utility performance and operations against regulatory requirements and industry best management practices (Kirmeyer et al. 2000, Friedman et al. 2005).

USEPA has expanded regulatory focus beyond maximum contaminant level (MCL)–based compliance, to incorporate risk-based approaches. An example of this change is provided in the Revised Total Coliform Rule, in which it is recommended that the total coliform nonacute maximum contaminant level be replaced with more detailed distribution system assessments. The Partnership's self-assessment process may provide a means for utilities to increase their familiarity with distribution system operations, building a framework that may prove beneficial in the case that a Revised Total Coliform Rule assessment is ever required in the future. Another example of this includes the Microbial/Disinfection By-product Rules, encompassing the Long Term 2 Enhanced Surface Water Treatment Rule and the Stage 2 Disinfectants and Disinfection By-products Rule (Stage 2 DBPR). Both rules require systems to determine an individual level of contaminant occurrence and identify the associated required treatment level and type. The Stage 2 DBPR requires systems to assess distribution system operations and contains a component (e.g., Operational Evaluation Levels) outlining an investigative framework aimed at prevention of ongoing disinfection by-product problems.

This shift toward assessment of distribution system water quality and associated operational and maintenance practices is reflected in the PSW optimization program, and the performance assessment described in Chapter 2. Systems that voluntarily take part in this program should realize significant benefits ensuring continued and future compliance with existing, proposed, and expected drinking water regulations.

There are several federal regulations that apply to distribution systems, a number of which are listed in Table 1-2. There are also additional local regulatory requirements that are not listed here. Each utility, including consecutive systems, must be attentive to all regulations that apply to its system. The list that follows is not comprehensive and does not include regulations for countries outside of the United States. PSW subscribers from countries outside the United States should be aware of and strive for compliance with all applicable regulations for their country and region. Although the focus of the Partnership is on optimization, the

Table 1-2.   USEPA distribution system regulations

Safe Drinking Water Act (and amendments)
National Interim Primary Drinking Water Regulations
Total Coliform Rule, Revised Total Coliform Rule
Surface Water Treatment Rule
Lead and Copper Rule
Information Collection Rule
Stage 1 Disinfectants and Disinfection By-products Rule
Interim Enhanced Surface Water Treatment Rule
Long Term 1 Enhanced Surface Water Treatment Rule
Stage 2 Disinfectants and Disinfection By-products Rule
Long Term 2 Enhanced Surface Water Treatment Rule
Unregulated Contaminant Monitoring Rule (UCMR) 2, UCMR 3
Water Security-related Directives and Laws
Secondary Maximum Contaminant Levels
Public Notification Regulations

self-assessment process should also improve utility ability to consistently comply with all applicable regulations.

Safeguarding Distribution System Water Quality

Public Health Consequences

Many regulations apply to water quality entering distribution systems (SDWA, USEPA). There are only a few that apply specifically to distribution systems (Total Coliform Rule, Surface Water Treatment Rule, Disinfectants/Disinfection By-product Rule, Lead and Copper Rule). Individually, these regulations have a specific focus and therefore a limited effect on overall distribution system water quality. These regulations were established to protect public health and, sometimes, improve tap water appearance and acceptability to users. Various regulated and unregulated contaminants have been identified in drinking water distribution systems. The health significance of many unregulated contaminants is yet unknown.

The US Centers for Disease Control and Prevention issues reports on waterborne outbreaks and, although the frequency of occurrence is declining, a percentage of outbreaks is associated with drinking water distribution systems. It is commonly understood that these outbreaks may be underestimated for various reasons. Although the causes vary, preventing waterborne outbreaks is a primary objective of water suppliers.

Ensuring Delivery of High-Quality Water

The goal of water service providers is to reliably deliver high-quality water to all users. This goal may not be achieved because of conditions in the water distribution network that are sometimes unintentionally degraded by operational actions. The objective of the PSW distribution system optimization program is to identify opportunities for improvement in system operations and to empower system operators with the knowledge to recognize and apply procedures that result in water quality improvement and enhance system reliability.

There are three primary causes responsible for deteriorating water quality in distribution systems: external intrusion, internal surfaces, and water chemistry (or microbiology). Several reports (Kirmeyer 2000, NRC 2006, Friedman et al. 2010) have determined that water quality and system reliability can be enhanced by preserving system integrity for: *water quality*, *hydraulic*, and *physical* properties.

Water treatment plants use a multiple-barrier approach to ensure high-quality water is continuously produced. Each barrier is optimized, which provides multiple obstacles to the passage of pathogens through the treatment system. Similarly, distribution system integrity requires optimization of three primary components, which are illustrated in Table 1-3. High-quality water and system reliability can be enhanced if all three integrity components are optimized.

Table 1-3. Optimizing multiple system components, treatment versus distribution

Distribution system integrity	Treatment plant integrity
Water Quality	Flocculation/Sedimentation
Hydraulic	Filtration
Physical	Disinfection

Preserving and maintaining distribution system integrity requires proper system design, understanding and application of operational optimization practices, and administrative support. There are many important factors that can be used to improve each system component and help prevent impacts that can negatively affect water quality and system reliability. However, there is a *primary* performance indicator that can be consistently applied to each major system integrity component. For the Partnership's treatment plant optimization program, the primary performance indicator for particulate removal is turbidity. The distribution system optimization program's performance indicators for water quality, hydraulic, and physical integrity are outlined in Table 1-4.

Partnership for Safe Water Optimization Philosophy

The Partnership uses optimization methods to improve system performance. The concept relies on beginning with a system that is "capable" of reliably delivering high-quality water when the operation is optimized. Optimization involves identifying performance limiting factors and designing and implementing an action plan that results in improved performance. System operators (and managers) must understand and apply the optimization concepts to achieve the goal of continuously delivering high-quality water to all users.

The areas in which performance-limiting factors have been broadly grouped (administration, maintenance, design, and operation) are all important because any one of these areas can individually cause poor performance. When completing the self-assessment process, the interrelationship of these categories, and the utility staff representing each category, to achieving the goal of reliably delivering high-quality water *must* also be carefully considered and understood.

Table 1-4. Distribution system integrity performance indicators

Category	Performance indicator
Water Quality	Disinfectant Residual (maintain adequate disinfectant residual)
Hydraulic	Pressure (maintain adequate pressure)
Physical	Main Breaks (reduce main break frequency)

The relationship of utility administration, design, maintenance, and operation in the delivery of high-quality water on a continuous basis is shown graphically in Figure 1-1. As shown, administration, design, and maintenance staff must all work together to provide the solid foundation of a system capable of delivering high-quality water.

It is the operation, or specifically the system operational control activities, that enables a capable system to produce high-quality water on a continuous basis. However, operators can easily become frustrated if attempting to optimize a system independently, without input and support of administrators, maintenance, and engineering personnel. Therefore, it is critical that the self-assessment be conducted via a team approach, including representative personnel from administration, maintenance, design, and operations at all levels of the organization. Key roles and responsibilities of each group must be established, and communication between all staff must be consistent and effective.

Distribution system optimization goals form the basis of discussions in several sections of the self-assessment. A capable system, as previously described, should be able to achieve the optimization goals. The intent of the self-assessment is that the water supplier will voluntarily set challenging goals for their system. This will foster implementation of an action plan that allows for the achievement

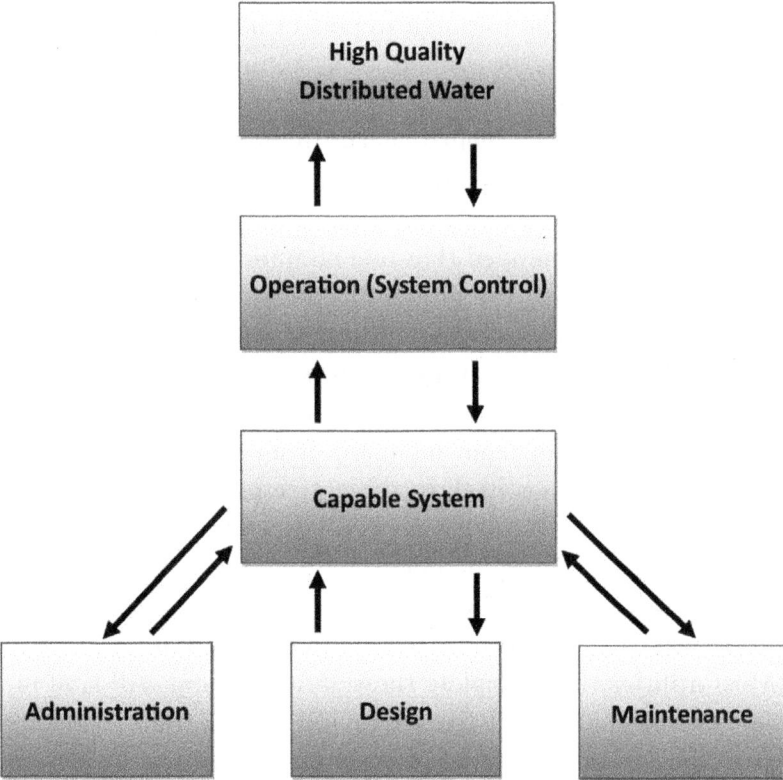

Figure 1-1. Capable system model

of goals set beyond regulatory requirements and results in the development of an optimized system over time.

Partnership for Safe Water Self-Assessment Approach

The self-assessment process is a collaborative team effort between management, operations, and maintenance staff. The team should develop and discuss the self-assessment schedule and assign members with appropriate knowledge, expertise, and responsibility to perform each task. The self-assessment is then performed by completing the analysis and self-assessment questions included in each chapter of the guide. There are no time requirements for the self-assessment. The process is self-paced, so that utilities may complete the process according to the schedule that provides them with the greatest amount of information and benefit, and the Partnership for Safe Water accepts self-assessment completion reports on a continuous basis.

One of the first assignments for the self-assessment team is to assemble the data and documents needed to complete the evaluation. Many of the optimization topics require the analysis of historical distribution system data, from operating records, for comparison with the program's optimization goals. Some utilities may not have sufficient data to complete portions of the assessment. In this case, action should be initiated to collect additional data, either prior to beginning that portion of the assessment or as an outcome of the self-assessment as an action item. The data evaluation process is used to determine the optimization status of the system and to identify performance limiting factors. A summary of the minimum data and documents needed to complete the self-assessment is listed in Table 1-8.

Generally, it is recommended that one chapter be addressed at a time, proceeding sequentially through the guide. However, if the team is large enough, the material in some chapters can be addressed simultaneously. Some chapters require consideration of assessment parameters that can cross utility department lines. It may be best to divide these tasks within a single chapter and assign them to the most suitable group of team members from multiple departments. These subgroups can report findings back to the entire team for full group consensus discussion, and, ultimately, these efforts can be combined for the final report.

The self-assessment topics included in this guide are not intended to be a comprehensive compendium of the *only* issues that distribution system operators should consider. These are the most common issues that most utilities should examine. Most utilities that complete the self-assessment will find more optimization issues (topics) that are of particular interest to their distribution system staff. It is common for some unique performance limiting factors to be identified that do not directly relate to a specific question included in this guide. Utilities should still discuss these topics and include them in their self-assessment, prioritize their importance, and develop an action plan to improve in these areas. There is no single correct way in which the self-assessment may be completed, although many utilities have chosen to implement regular, recurring meetings of utility staff in which the assessment and optimization principles are discussed.

Although utilities may complete the self-assessment in any manner considered to be appropriate, the self-assessment should be a team effort among management, operations, maintenance, and laboratory staff. The greatest learning and benefit can be obtained by involving staff at all levels of the organization in the self-assessment process and making continuous improvement and optimization a part of everyone's function at the utility, regardless of where a position may reside on the utility's organizational chart. Some suggestions for encouraging broad team involvement in the process include:

- Seek management support for completing the self-assessment.

- Formally establish a self-assessment team that includes representatives from across the utility's distribution system staff. The composition of this team may vary depending on utility size and structure, but it is recommended that the team involve staff from as many levels and functions of the utility as possible.

- Establish regular self-assessment meetings during which staff is encouraged to discuss distribution system performance as well as specific self-assessment questions. The frequency of these meetings may vary, but many utilities report scheduling team meetings at a frequency of once weekly to once monthly.

- Act on team findings: if an action item is identified that is able to be, or should be, addressed now, do not wait until the self-assessment is completed to take action. Allow operators to take ownership of actions they identify, including conducting special studies or development of operational tools, to start taking steps toward optimization. A report of progress made can be included in the self-assessment completion report.

- Be mindful that the purpose of the self-assessment is not a fault-finding mission but rather a systematic process to identify issues that are negatively impacting distribution system performance and developing solutions.

There are many more ways that utilities can encourage and promote team involvement in the self-assessment process, limited only by the ideas developed by utility staff. If a utility finds that it cannot complete the self-assessment as a broad team composed of utility staff, it may consider how these attitudes may limit performance, particularly in the administrative section of the self-assessment. The utility may find that working through the self-assessment process as a team results in improved communication and understanding at all levels.

As previously stated, the self-assessment process is also a self-paced process. There is no time limit imposed during which a utility is required to complete the self-assessment. The schedule for completion of the self-assessment is set by the utility in a manner that allows utility staff to derive the greatest benefit from the self-assessment process, while working within the resources available.

Guide Layout

The structure of this guide parallels the major steps of the self-assessment and provides the complete framework required to complete a self-assessment of distribution system performance and operations according to the PSW guidelines.

This guide is designed to be used in conjunction with additional Partnership tools to assist staff in conducting a comprehensive self-assessment of distribution system operations and performance and preparing a self-assessment completion report. The self-assessment principles and questions may be applied to any type of distribution system, including municipal retail systems, wholesale systems, and consecutive systems. Utility staff needs only address the self-assessment questions that are relevant to their system (for example, a wholesale provider may not be required to address questions related to fire hydrants if it does not own any). Because of the nature of the self-assessment questions, it is recommended that systems complete the self-assessment independently to derive the maximum benefit from the process.

The guide is organized to step utility staff through the process of completing the self-assessment in a logical fashion. The self-assessment process addresses the following five categories:

- Chapter 2—Performance Assessment
  - ▲ Disinfectant residual
  - ▲ Pressure management
  - ▲ Main breaks
- Chapter 3—Operational Performance Improvement Variables
- Chapter 4—Design Evaluation
- Chapter 5—Application of Operational Concepts
  - ▲ Process control testing and operator application of concepts
- Chapter 6—Administration
  - ▲ Evaluation of administrative policies, staffing, and funding

These chapters are followed by chapter 7, which covers the topics of prioritization of performance-limiting factors and development of an action plan. The guide's Appendix contains a variety of resource materials that are relevant to utilities completing the self-assessment process. A more detailed description of the contents of each chapter follows.

Chapter 1 provides background information about the self-assessment process, how to begin the process by establishing a team, and descriptions of potential strategies for completion of the self-assessment process.

Chapter 2 describes how to conduct a performance assessment to determine the existing level of performance of the utility's distribution system, with respect to the primary optimization parameters of disinfectant residual, pressure, and main breaks. Based on the findings of the performance assessment, utility staff will understand how distribution system performance compares with the Partnership's or utility's internally established optimization goals.

After completion of the performance assessment, the self-assessment team should then progress to complete chapters 3 through 6. These chapters examine distribution system Performance Improvement Variables that all relate to the program's key optimization parameters, as well as design, operational, and administrative areas to help utilities identify any performance-limiting factors that may inhibit optimized distribution system performance. Each chapter contains a list of self-assessment questions for the team's consideration. At the end of each chapter is a table that summarizes responses to each question, including columns to indicate if each factor considered is Optimized, Partially Optimized, or Not Optimized. If the team considers an area to be optimized, the factor listed is not considered to be limiting distribution system performance and should not be considered further. The self-assessment completion report should contain documentation or an explanation that supports the selected optimization status, which may include information such as that generated by a special study, in some cases. However, if the status of a particular item is determined to be Partially Optimized or Not Optimized, the factor is considered to be limiting optimized distribution system performance and should be considered in the identification and prioritization activities of chapter 7. Likewise, the self-assessment report should include documentation to support the conclusion. Ultimately, factors that are determined to not be optimized should be associated with an action-improvement plan, as described in chapter 7.

Chapter 7 presents a method for the self-assessment team to prioritize the most important factors that may be limiting distribution system performance, as determined from the assessment conducted in the previous chapters, in combination with the prioritization ranking calculated using the Partnership's Optimization Assessment Tool software (provided to PSW subscribers). This provides clarity to the team as to where follow-up activities should be implemented to improve distribution system performance. The goal of the Partnership's self-assessment process is to optimize existing distribution system assets and operations without major capital improvements. This may be a challenging goal for distribution systems. If the outcome of the self-assessment process indicates that capital improvements are required to achieve performance improvements, the action planning process can help utility staff to develop a systematic approach to capital improvement planning that enables them to complete improvements according to a realistic and achievable schedule. Utilities do not need to wait until the self-assessment is completed to begin addressing action items. Utility staff may find it beneficial to begin work on action items while the self-assessment is in process. Any progress made on action items should be documented in the self-assessment completion report.

Each of the self-assessment topics is presented using the same format (Figure 1-2), broken down into sections titled Understanding, Status, and Action. The Understanding section discusses key issues related to the identified topic and presents reasons they are critical to meeting the distribution system optimization goals. The Status section contains self-assessment questions, related to the topic, that the utility will address to determine their optimization status. By addressing the self-assessment questions, utility staff will gain insight into how its current practices support the optimization goals. By comparing the information on current practices with that discussed in the Understanding section, utility staff can identify potential factors limiting optimized performance. The Action section contains information that may be used to develop action plans, designed to address any deficiencies identified while completing the self-assessment questions and improve performance.

The key element of the self-assessment is answering the questions included in the Status section for each topic. The utility's self-assessment team should

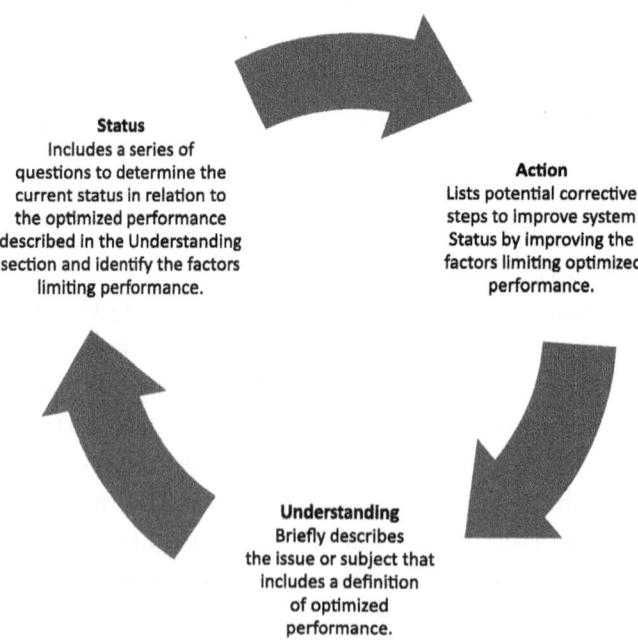

Figure 1-2.   Self-assessment approach

consider its responses carefully. Sometimes, additional performance data must be gathered and evaluated to compare with the topics being addressed in each question. In other cases, current procedures or policies may adequately address the question.

A summary of the self-assessment questions is provided at the end of each chapter. The self-assessment team should have discussed each self-assessment question and determined the utility's optimization status, after which the responses can be recorded in the summary table and in the Optimization Assessment Tool software, provided to PSW subscribers. Responses to self-assessment questions can indicate that the topic is Optimized and Documented, Partially Optimized, or Not Optimized. An Optimized and Documented response to a question may be satisfactory if documentation is available to support this response. The summary tables (example shown in Table 1-5) list some questions that should include a written explanation, even if the response is Optimized and Documented. These written explanations may be brief but are highly recommended to include *numerical values* for comparison with any benchmark in the question. A Partially Optimized response is often caused by a multipart question, particularly if some areas addressed in the question are fully optimized while the utility may be working to achieve optimization in other areas addressed in the question. The Partnership provides tools to assist utilities with responding to self-assessment questions and creating a self-assessment completion report. The Distribution System Self-Assessment Example Report provides examples of effective responses to self-assessment questions and the types of information that may be included to support a response of Optimized and Documented. The Self-Assessment Completion Report Template is another tool that provides utilities with a structure for creating the self-assessment report and developing action plans.

The self-assessment topics that are considered to have a status of Not Optimized or Partially Optimized are prioritized using the method described in the continuous improvement plan chapter (chapter 7). An action plan is developed to optimize factors that hold the most promise for improving system performance. The results of these efforts are monitored and reported annually, along with disinfectant residual, pressure, and main break data, to the Partnership as part of the program's annual reporting process. The system performance criteria are examined at least annually to determine if the optimization efforts are effective. Depending on the results, adjustments may be needed to further optimize system performance.

Distribution System Optimization Goals

Goal setting is an important principle in the PSW program. The following sections describe principles of goal setting and help to define performance goals relative to the program's three main optimization areas of disinfectant residual, pressure management, and main break frequency. A more detailed discussion

Table 1-5.   Example of a self-assessment performance-limiting factors summary

Self-assessment category	Questions for gauging optimization status	Optimization status			Comments
		Optimized and documented	Partially optimized	Not optimized	
Master Plan	Does the system have a 20-year (or longer) plan for improvement and expansion of the distribution system?			✓	Narrative responses for all questions should be included in the self-assessment completion report.
Capital Improvement Program	Is there a current (updated) short-range improvement plan that is designed to meet current needs and system deficiencies?		✓		

of these goals is included in chapter 2. Utilities are also encouraged to consider the benefits of establishing utility-specific goals for parameters that may not be specifically addressed in this guide. Goals are important because they provide a benchmark relative to which utilities may assess performance and provide a standard of performance for which to strive.

Setting Goals

What is a goal? A goal is a desired result or a destination. A goal is a specific objective for which to strive. The process of setting goals is important for water distribution staff, because assessment of performance in comparison to goals indicates the current status of distribution system operations, and assessment of performance at set intervals allows the progress of distribution system performance toward these goals to be measured. Goals can be set for any aspect of distribution system operations and performance. The primary goals of the Partnership's distribution system optimization program focus on the three primary indicators of distribution system water quality, hydraulic, and physical integrity, as outlined previously in Table 1-4.

The PSW's distribution system optimization goals are outlined below for each of the primary optimization parameters. Additional details regarding each optimization parameter are included in chapter 2. In addition to these goals, staff may also wish to develop their own system-specific goals for areas not directly addressed in this guide. Setting goals, for any distribution system process, that are challenging, but still attainable, can be an important means of engaging staff and keeping levels of motivation high throughout the self-assessment process.

Performance goals for many optimization areas are listed in this self-assessment guide. These are provided to help system operators compare their operation with measures of excellence. It is likely that systems will need to make improvements before consistently achieving *all* the goals offered in this guide. Distribution systems are not required to meet the program's optimization goals to complete the self-assessment process or participate in the program. Systems should be committed to accurately and honestly assessing their performance relative to these goals and implementing actions to improve and optimize where it is indicated.

Disinfectant Residual

PSW subscribers apply secondary disinfection and/or manage distribution systems that convey water with disinfectant residuals as a requirement of program participation. Optimized systems meet the distribution system disinfectant residual optimization goals as displayed in Table 1-6.

These goals listed in Table 1-6 are for 95% of the routine distribution system samples taken each month from optimized sampling locations. Individual routine sample sites also should not have consecutive (prescheduled) disinfectant residual concentration measurements fall outside the performance goals. Additionally, optimized systems include distribution system sample locations that target problem sites and all storage tanks, as described in greater detail in chapter 2. Optimized distribution systems are in compliance with local regulations for disinfection by-products.

Pressure Management

A distribution system optimized with respect to pressure management should maintain a minimum pressure of no less than 20 psi. This goal should be met under normal operating conditions, including maximum daily demand and fire flow conditions (excluding emergencies). This goal is for 99.5% of the minimum daily readings from permanent pressure sensors located at sites of minimum pressure. A minimum pressure of 35 psi or greater should be maintained as a monthly average of daily minimum values. Emergency conditions, such as main breaks and power outages, require implementation of preapproved procedures to protect the public health. The maximum pressure range (difference between the

Table 1-6.   Disinfectant residual optimization goals

Disinfectant parameter	Optimization goal (95% of measurements)
Free chlorine	≥ 0.20 mg/L and ≤ 4.0 mg/L
Total chlorine (chloramines)	≥ 0.50 mg/L and ≤ 4.0 mg/L
Chlorine dioxide	≥ 0.20 and ≤ 0.80 mg/L
Disinfection by-products	In compliance with local regulatory requirements

minimum and maximum within a pressure zone) should not be greater than a predetermined utility specific goal, for 95% of measurements. The system's maximum pressure should not be greater than a predetermined utility specific goal in 95% of measurements. Flexibility in the development of utility-specific goals for pressure fluctuations and maximum pressure is allowed due to system to system variability in pressure conditions. The Partnership's optimization goals for pressure are summarized in Table 1-7. Additional details pertaining to pressure monitoring are included in Chapter 2.

Main Break Frequency

The optimization goal for main break frequency is an annual maximum of 15 recorded breaks and leaks for every 100 miles of utility-owned distribution pipeline. This main break frequency goal may vary in its difficulty to achieve from one utility to another. It may take many years of optimization efforts for some utilities to achieve this optimization goal. A reduction in the main break frequency over time (rolling 5-year trend) is an indication of progress toward optimized performance and may be used as an alternate optimization criteria. The main break frequency goal does not include breaks detected as part of the utility's leak detection efforts. As further discussed in chapter 3, an active leak detection program is one component of a distribution system optimized for water loss control. Chapter 3 also provides additional information regarding the definition of a main break, as applied for performing the Partnership's data collection and evaluation process.

Establishing, and striving to meet, these optimization goals is a prudent approach for distribution systems, given the ongoing focus on establishing future distribution system regulations. The Partnership, however, recognizes that these goals may be challenging for every water system to attain in the short term. Therefore, some systems may wish to set suitable interim goals in their performance assessment that challenge the system toward significant improvements, as they work toward achieving the program's optimization goals. This is the purpose of the Partnership program, and it is the intent of the self-assessment to provide a structured procedure for systems to assess and improve performance.

Table 1-7.   Pressure management optimization goals

Pressure measurement	Optimization goal
Minimum pressure	20 psi for 99.5% of minimum daily measurements
Maximum pressure	Less than utility-set maximum for 95% of measurements
Pressure fluctuation range	Less than utility-set pressure fluctuation range for 95% of measurements

Self-Assessment Data Requirements

Obtaining the most accurate and relevant distribution system self-assessment results is dependent on several factors, including engagement of a representative self-assessment team (as previously discussed) and the availability of sufficient and accurate distribution system performance, operational, statistical, and administrative data. A list of the minimum data required to complete the self-assessment process is displayed in Table 1-8. Note that all of these data are not reported to the PSW in raw form; however, the data included in Table 1-8 will be helpful for utility staff when addressing the self-assessment questions included in this guide.

Utilities that do not have all of these data available will still typically be able to complete the self-assessment process. Any data deficiencies identified during the self-assessment process should be addressed in the utility's action plan. If a need for additional data is identified, utility staff should not hesitate to act to address the issue, even while still completing the self-assessment process. A significant data deficiency (for example, a utility lacking any distribution system disinfectant residual data) may indicate the need to delay completion of the self-assessment process until sufficient data can be collected to complete the performance evaluation.

Utility data may often reside in a variety of storage locations, in both electronic and hard copy form. Additionally, databases may offer limited access to select utility staff with a defined objective for accessing the data. Staff may not always be aware of the full range of distribution system data that is collected and stored by the utility. In some cases, it can be helpful to engage a member of the utility's information technology staff on the self-assessment team to help team members to identify, locate, and correlate the components of the distribution system data required to complete the self-assessment process.

Getting Started With the Self-Assessment Process

The following steps indicate a recommended approach for getting started with, and completing, the self-assessment process.

Step 1—Complete Chapter 1—Conducting the Self-Assessment

Assemble a self-assessment team with subject matter experts from all distribution system stakeholders: operators, management, water quality, engineering, consultants, contractors, maintenance, information technology, and customer service. Organize a self-assessment plan and proposed timeline that involves data gathering, performance assessments, report writing, data analysis, and create a completion schedule that is attainable, while realizing that the self-assessment is a self-paced process. Conduct periodic meetings of the team to assess progress and adjust to the completion schedule. Upon completion of the self-assessment, the

Table 1-8. Minimum data requirements to complete the self-assessment

Number	Data description	Guide reference
1	Minimum daily disinfectant residual for 12 months at distribution system routine sample locations (location, value), storage tanks, and entry points to the distribution system, reported using the data collection software provided by the PSW. A minimum 2 years of disinfectant residual data is required for self-assessment report submission (baseline year and current year).	C2* (Disinfectant Residual)
2	All total trihalomethane (TTHM) and haloacetic acid (HAA5) routine test results for 12 months (date, value, location), reported using the data collection software provided by the PSW. Annual disinfection by-product trending indicating running annual average. A minimum 2 years of disinfection by-product data is required for self-assessment report submission (baseline year and current year).	C2 (Disinfectant Residual), C3 (Disinfection By-Products)
3	Daily minimum pressure readings from permanent sensors for the most recent 12 months, reported using the data collection software provided by the PSW. Ideally, data are collected from sensors located at the low- and high-pressure locations in each pressure zone.	C2 (Pressure Management)
4	Main break records for several years (location, date, number, condition), reported using the data collection software provided by the PSW. A minimum of 5 years of data is required for demonstration of a declining 5-year frequency trend, whereas a 10-year duration is considered idea.	C2 (Main Break Frequency)
5	Records of the annual number of technical water quality complaints and the number of utility accounts.	C3 (Customer Complaints)
6	Flushing velocity and before/after disinfectant residual data for flushing procedures that are initiated to address low disinfectant residuals.	C3 (Flushing)
7	Total number of valves and hydrants in the system, valve and hydrant exercise and inspection records, and a summary of the number of valves and hydrants exercised annually.	C3 (Maintaining Hydrants, Valves, and Blowoffs), C4 (Valves and Hydrants)
8	Hydrant repair records and calculated time to restore service after detection.	C3 (Maintaining Hydrants, Valves, and Blowoffs), C4 (Valves and Hydrants)
9	Internal corrosion testing practices, including analysis parameters and the number of tests performed annually.	C3 (Internal Corrosion Control)
10	Summary of analytical results for free ammonia, nitrate, nitrite, and other nitrification indicator parameters for distribution systems using chloramines.	C3 (Nitrification)

*C2 indicates that this item is discussed in chapter 2. Guide references noted may not represent the only locations at which the topic is addressed in the guide.

Number	Data description	Guide reference
11	Pipeline renewal and replacement rate, including the annual miles of pipeline replaced and the total miles of pipe in the entire distribution system. Specify the miles of unlined metal pipe and the miles of such pipe replaced, as well as miles of pipe more than 75 years old.	C3 (Pipeline Installation, Rehabilitation, and Replacement), C4 (Pipeline Materials)
12	Storage tank cleaning records that indicate the frequency of cleaning and any observations.	C3 (Storage Facility Operation and Maintenance), C4 (Storage Facilities Materials and Construction)
13	Water age records at key sites demonstrating the annual maximum water age. If a hydraulic model is used, water age predicted by the hydraulic model.	C3 (Water Age, Modeling)
14	Calculation of the volume of annual real losses, real losses expressed as gallons/service connection/day, and infrastructure leakage index (ILI) using the AWWA/International Water Association (IWA) water audit method.	C3 (Water Loss Control)
15	Distribution system schematic (map)	C4 (Asset Management)
16	Asset inventory	C4 (Asset Management)
17	Distribution system pipeline type inventory and installation dates	C4 (Pipeline Materials)
18	Storage facility type and installation dates	C4 (Storage Facilities Materials and Construction)
19	Pump type, size, and installation dates	C4 (Pumping Facilities)
20	Valve and hydrant number and installation dates	C4 (Valves and Hydrants)
21	Determination of the following benchmarks (reporting is optional): (*Debt Ratio, O&M Cost per Account, System Renewal Rate, Training Hours per Employee, Interest % of Budget*).	C6 (Administration)
22	Summary of field testing performed for chemical and biological parameters, the methods/instrumentation used, and numerical goals	C5 (Operations)
23	Standard Operating Procedures (SOP) related to distribution system operation. The self-assessment presents a good opportunity for review and revision of the utility's SOPs.	All

team presents a self-assessment completion report to the Partnership for peer review. After receiving recognition for completion of the self-assessment, the team meets periodically to review system performance and to prepare annual data reports for submission to the Partnership program.

Step 2—Complete Chapter 2—Performance Assessment

Determine if the current levels of performance are meeting the optimization goals by comparing distribution system data with the optimization criteria for water quality integrity, hydraulic integrity, and physical integrity indicators. Review of the completeness and accuracy of the data should be considered part of this assessment. The program's optimization goals are not all-encompassing; therefore, system managers should complete the entire self-assessment to decide how performance can be further improved even if the goals are met. For systems not meeting the goals, self-assessment questions can aid in identifying the potential factors limiting optimized performance so that an action plan may be developed and implemented to achieve improved performance. If the performance assessment indicates that the optimization goals are achieved, utility staff needs to ensure that utility performance is not the result of a relatively new system that masks the lack of optimization skills and leads to complacency. Also, utilities may want to adopt the self-assessment process as a part of a continuous quality improvement effort (for example, complete appropriate portions of the self-assessment every 3-5 years).

Step 3—Complete Chapter 3—Assess Status of Performance Improvement Variables

Many factors contribute to optimized distribution system performance. Chapter 3 contains self-assessment questions for Performance Improvement Variables that can affect distribution system performance. Optimizing each of these processes or procedures results in an optimized distribution system. Determine the optimization status for each of the Performance Improvement Variables by answering the self-assessment questions included in this section of the guide. Comparing the current status with optimized goals can reveal performance limiting factors and opportunities for improvement.

Step 4—Complete Chapter 4—Design

Determine if there are aspects of system design that may limit optimized performance. The self-assessment approach selectively identifies exceptions in system design that are known to contribute to water quality deterioration. Examples are areas served through unlined cast iron pipe, or the use of materials in contact with drinking water that are not certified for potable use. This chapter also addresses the use of distribution system hydraulic models and asset management programs.

Step 5—Complete Chapter 5—Evaluate Operator Application of Concepts

Determine if there are operational practices that may be limiting optimized performance. The operations topics assess the system's operational control program and the operations staff's ability to interpret, react to, and communicate water quality changes. Staff must have the knowledge and authority to identify and communicate performance deviations and apply correct controls to achieve the performance goals. Standard operating procedures are a key component of the operational concepts covered in this portion of the guide. *Topics included in this step are most often identified as an area needing improvement.*

Step 6—Complete Chapter 6—Perform Administrative Assessment

Determine whether utility administration provides the support needed to achieve optimized system performance. These administrative topics assess policies, staffing levels, training, engagement, and funding required to support a capable system and allow utility staff to achieve the desired performance goals. Financial benchmarks may be considered in this assessment, although reporting of specific financial benchmarks is optional. Topics included in this area have proven to be the most difficult to assess objectively.

Step 7—Complete Chapter 7—Create and Implement an Action Plan for Improvement

Begin by summarizing responses to self-assessment questions, included throughout the guide, that are identified as Partially Optimized or Not Optimized. These are grouped by optimization category to simplify the evaluation. Assemble and prioritize a list of factors and categories that can improve performance, and identify activities that will lead to optimized performance. The Partnership's Optimization Assessment Tool software is a resource that helps guide utilities through this process. It is important to prioritize the factors according to their estimated impact and urgency to improve performance. Use these factors to develop an Action Plan to improve performance. The plan should include both long- and short-term actions to address the performance limiting factors identified by utility staff. Apply utility resources to the highest ranking factors first and to lower ranking factors as time and resources allow. *Typically, performance improvements often have to be addressed together because an interdependent combination of factors typically contributes to overall system performance.*

Self-Asssessment Completion Report

The Partnership issues recognition awards to subscriber utilities that complete the self-assessment process and submit a self-assessment completion report that successfully completes the peer-review process. The self-assessment process is outlined in this document. The Partnership provides guidelines and templates for preparing a self-assessment completion report and describes the procedures for report submission. The self-assessment completion report is reviewed by utility volunteers composing the Program Effectiveness Assessment Committee. These utility volunteers are subject matter experts in drinking water utility optimization and review the report based on specific parameters to determine that a good faith effort has been made to assess distribution system operations and performance. These guidelines and procedures specific to the distribution system self-assessment process are available on the PSW program Web pages (**www.awwa. org/partnership**). The evaluation parameters used in the peer review of the self-assessment completion report are included in the Appendix.

The optimization process does not end after completion of the self-assessment, and, for Partnership subscribers, submission of the self-assessment completion report. The optimization process is ongoing as distribution system operations, and the distribution system itself, continues to evolve and change with the passage of time. Once action items are identified and prioritized, utility staff should work on implementing and completing these actions. On a regular basis, utilities should continue to review distribution system performance for disinfectant residual, pressure, main breaks, and/or other parameters to ensure that performance continues to be maintained or improved. Systems that were performing at a very high level before completing the self-assessment process may not see improvements reflected as significantly in the data and should look to other areas to measure improvement, such as reliability, consistency, administration, and operator knowledge. Utilities should also plan to review progress made relative to the action plan developed and continue to identify any additional areas that may limit optimized performance. Utilities may even wish to repeat the self-assessment process, or applicable areas of the evaluation, on a regular basis, such as every 3-5 years. If needed, new action plans can be developed to address any newly identified performance limiting factors to ensure that the distribution system remains on track to continuously improve. By implementing such a process at a utility, the self-assessment and continuous improvement processes can become important components of utility culture.

CHAPTER 2

PERFORMANCE ASSESSMENT

Introduction

Optimized distribution system operation is achieved by optimizing the integrity of the three categories (Friedman et al. 2010) that protect the water from deterioration and enhance system reliability: water quality, hydraulic, and physical. Optimization criteria for the most important indicators in each of these areas are used to evaluate the performance of the system, as illustrated in Table 2-1.

Meeting the optimization goals for each criterion suggests optimized performance may be attained. However, these are indicators only and do not fully describe the operation in all areas. The performance assessment is one important step in an overall optimization process that includes a thorough examination of many important aspects of distribution system operation.

The objective of the performance assessment is to collect specific distribution system data associated with each of the three optimization criteria (disinfectant residual, pressure management, and main break frequency) and compare this information with numeric optimization goals. The performance assessment uses data and equipment that many utilities already have, including disinfectant residual data from routine monitoring sites, pressure data from existing pressure sensors, and main break statistics from historical records. There is a separate subsection discussion and assessment software (provided to Partnership for Safe Water [PSW] subscribers) for each of the three criteria. Besides the general headings included in this chapter (understanding, status, action,

Table 2-1.   Distribution system optimization criteria

System integrity category	Optimization criteria
Water Quality	Disinfectant Residual
Hydraulic	Pressure Management
Physical	Main Break Frequency

and performance-limiting factors) the discussion of each optimization criteria includes many of the following additional topics:

- Numeric performance goals
- Data requirements
- Optimized monitoring locations
- Testing frequency
- Method of measurement
- Analytical requirements
- Data analysis
- Description of full optimization

The performance assessment for these three criteria can be completed using data and equipment already in use by most utilities. Utilities using available data that do not meet the optimization requirements, or that are not collected in a sufficient quantity to meet the optimization requirements, should include improving data collection procedures and/or methods as one of the utility's performance-limiting factors to be addressed in the action plan. The distribution system cannot be considered fully optimized until sufficient data are collected and evaluated to provide the ability to assess the system's optimization status.

Distribution system operators determine the optimization status for each of the criteria separately. System optimization cannot be achieved unless the performance goals for all three optimization parameters are met, using sampling and analysis techniques performed according to the optimized data collection requirements. The numeric performance goals are challenging because they are meant to represent the performance of an optimized distribution system.

PSW subscribers are provided with software to assist with collecting and summarizing distribution system performance data. Detailed instructions are included with the software programs provided to PSW subscriber utilities. Participating utilities should read through this guidance manual to become familiar with the self-assessment requirements, and review the software guides, to familiarize staff with use of the performance assessment software before beginning data entry. This will result in the most efficient use of time and effort committed to this process.

DISINFECTANT RESIDUAL

Disinfectant residual has been identified as a useful indicator of distribution system water quality integrity (USEPA 1989a-b, 1998a-b, 2002b, 2006a-c, 2007; Kirmeyer et al., 2000; AWWA 2005; Friedman et al. 2005; NASTB 2006; Routt et al., 2009; ANSI/AWWA C652-11; AWWA G200-15;). An extensive literature search, utility questionnaires, utility workgroups, and expert workshop deliberations resulted in identification of disinfectant residual as the most often used and widely cited water quality indicator for assessment of distribution systems (Friedman et al. 2010). This is likely because of its use as one of the final treatment barriers, its interaction with various water impurities, its prevalence and longevity of use, and ease of measurement. Disinfectant residuals alone may not be protective, but can serve as an indicator of water quality in distribution systems (NASTB 2006, USEPA 2006a-b).

Understanding

Free chlorine, chloramine, and chlorine dioxide are approved as secondary disinfectants under federal drinking water regulations in the United States and in many other countries (USEPA 1989a, 1998b). Chlorine and chloramine are the most commonly used residual disinfectants (AWWA Disinfection Systems Committee 2008a-b). Therefore, free and total chlorine disinfectant residuals are emphasized in this performance data assessment and optimization process. Effective use of disinfectant residual for system optimization requires an understanding of concepts related to chlorine and chloramine chemistry, applications, monitoring, and interactions within distribution systems (AWWA 2006a, 2013c). The content provided in this guide is largely drawn from the Water Research Foundation project #4109 final report (Friedman et al. 2010). However, additional content and references are cited throughout this guide.

Maintaining adequate chlorine residual throughout the distribution system is integral to ensuring drinking water quality. Commonly, local and federal regulatory requirements exist for disinfectant residual, applicable to surface water systems and many groundwater systems. Although disinfection waivers exist in some areas, they are becoming increasingly rare, and the maintenance of a residual disinfectant is a requirement for Partnership participation. Detailed explanations of complex chlorine and chloramine chemistry and related concepts are beyond the scope of this guide. Users are encouraged to consult the cited references and to research other sources, as needed, for further understanding (White 1999; AWWA 2006a, 2012c, 2013b).

The disinfectant residual optimization process and data collection software were developed for systems that practice secondary disinfection using free chlorine, total chlorine (chloramine), or chlorine dioxide. Blended systems (systems using both free chlorine and chloramine in different parts of the system) and systems that vary the type of disinfectant used throughout the year (for example, a free chlorine "burn" for chloraminated systems) may also successfully use the software and complete the self-assessment process. Utilities that do not use secondary disinfection in all or parts of their distribution systems are not eligible to participate formally in the PSW. However, using the disinfectant residual optimization tools to track system performance, even in partially chlorinated systems, should be useful to demonstrate and improve system water quality performance.

The disinfectant residual data collection software can accommodate free chlorine, total chlorine (chloramine), chlorine dioxide, or combinations of these disinfectants. When combination residual types are used in the same system, utilities will need to control and document this practice. In this case, notes should be added to the disinfectant residual data, and explanations should be included in the self-assessment report. For systems that switch disinfectants throughout the year, for example those that may apply a free chlorine "burn," it is recommended to use separate spreadsheets for the collection of free chlorine data versus total chlorine data because of differences in the performance goals for each disinfectant. It is also recommended that systems using multiple disinfectants throughout the system use separate spreadsheets for the collection and reporting of data for each disinfectant.

Utilities with large systems and extensive disinfectant residual monitoring programs and data will need to develop plans to compile appropriately selected data from multiple sources and at optimized sampling locations. Documentation is important. The self-assessment completion report should describe any special studies and intensive monitoring programs that can add to increased understanding of system performance. Disinfection by-product (DBP) results are included in the disinfectant residual performance data assessment and are an important aspect of disinfectant residual performance. Besides meeting all applicable regulations for this parameter (USEPA 1998b, 2006a, or appropriate local regulations), optimized systems implement methods to reduce DBP formation. Because trihalomethanes have a tendency to increase in concentration over time, DBPs may be of particular concern to utilities with extended distribution systems that may experience high residence times.

Disinfectant residual is used to demonstrate distribution system optimization for water quality integrity. The utility tracks and trends systemwide disinfectant residual levels and compares results from representative points in the distribution system, including as many optimized sampling locations as possible, to gain insight about how distribution system infrastructure and operations impact disinfectant residuals.

Site-specific disinfectant residual data trending can demonstrate optimization or track improvement as utilities identify and implement changes. Compliance with applicable regulations for disinfectants, DBPs, coliform bacteria, and corrosion control must be maintained. Also, systems must work to ensure there are no persistent aesthetic concerns with the potential to mask overall water quality performance or performance trends. At times, these competing regulatory requirements may appear to be in conflict with one another. However, it is important to achieve and maintain compliance in all areas, a concept commonly referred to as "simultaneous compliance." To maintain compliance in all areas, it is important that the potential impact of treatment and/or distribution system modifications be considered as broadly as possible.

Review of data from sample sites across the system will likely reveal patterns. Data patterns may be more readily identified by mapping data for chlorine residual and other parameters, in addition to examining graphical or tabular data. Added monitoring and data trending for comparison with parameters such as water age, pH, temperature, DBPs, and heterotrophic plate count (HPC) can be helpful to discover whether an observation is site-specific or systematic. Possible site or area-specific causes of residual loss may include:

- Cross-connections
- Contaminant intrusion from leaks or breaks
- Poor-quality main installations
- High water age
- Degradation from pipe material and corrosion by-product buildup
- Sediment or biofilm accumulation
- Nitrification
- Poorly controlled blending of free chlorine and chloraminated waters

Aesthetic problems may mask other water quality performance issues and cause disinfectant residual-based optimization results to be nonrepresentative. Examples include the normal presence of high levels of oxidized manganese in distributed water, which can interfere with disinfectant residual measurements (AWWA 2012c). If oxidized manganese is a concern, methods exist for overcoming this interference with the DPD chlorine analysis chemistry and with amperometric titration. Consult the instrument manufacturer or *Standard Methods for the Examination of Water and Wastewater* (current edition) for additional information. Another example could be systems with normally occurring odors, causing interference with the recognition of chlorine odors (Kirmeyer et al. 2000, NASTB 2006).

Booster chlorination is the practice of adding chlorine directly in the distribution system to raise residuals in a limited area of the system, such as a remote area or downstream of a storage facility. Boosting of residuals in the system can provide a more evenly distributed chlorine residual by applying it only where it is needed versus raising the concentrations fed to the entire system to maintain residuals in remote areas. This practice can also help to reduce system DBP concentrations. Chlorine booster systems must be carefully set up and controlled to meet site safety, dosage, and monitoring needs. Continuous online analyzers can be informative for remote surveillance and control of booster chlorination facilities if properly calibrated and maintained. Systems that feed chlorine and ammonia to boost chloramine residual should pay close attention to the chlorine to ammonia ratio, both before and after booster chloramination, to ensure that an optimized ratio is maintained. In chloraminated systems, booster chloramination may involve the addition of both chlorine and ammonia to maintain the optimal chlorine to ammonia ratio to avoid nitrification or other system challenges.

Because of the nature of chlorine breakpoint chemistry, blending of free and chloraminated waters, unless precisely monitored and controlled, has the potential to result in residual loss or undesirable odors. Systems converting between chloramine and free chlorine residuals for seasonal nitrification control or initial startup should monitor free and total chlorine, and other associated chloramination parameters, and conduct flushing, as needed, to accelerate the changeover between the residual types.

Chloramine residuals undergo auto decomposition over time, releasing ammonia, which can serve as a nutrient for nitrifying bacteria. Once established, nitrification can be difficult to remediate and can cause many water quality problems, as described in the nitrification section of chapter 3. Nitrification can be problematic in high water-age areas such as storage facilities and low-flow mains, particularly during warmer seasons. Therefore, utilities using chloramine should set up a program of monitoring and responding to nitrification indicator parameters (for example, nitrite, free ammonia, heterotrophic plate count (HPC), and rapid chloramine loss) to help identify areas prone to stagnation and improve operations in these areas. It is recommended that systems develop utility-specific nitrification action plans that provide specific guidance about the appropriate actions that should be taken under various water quality conditions relating to nitrification.

Chlorine residual measurements taken from storage facilities should be collected routinely. Sample collection should be timed to ensure that sampling is conducted when water is leaving the tank. In some instances, storage facilities can be practical locations for the application of continuous chlorine analyzers. If tank stratification is suspected, measuring chlorine residual at a variety of depths within a storage tank, potentially conducted as a special study, may provide information helpful to learning about storage tank performance.

Chlorine is highly toxic to aquatic life. This calls for special consideration where water from the system must be discharged to the environment (such as system flushing). Low levels of free chlorine residuals will dissipate more quickly when discharged, but high free chlorine residuals or low-level chloramine residuals may need to be removed, for example through reaction with a dechlorinating agent, such as vitamin C (ascorbic acid) or sodium thiosulfate. Utilities should be aware of local regulatory requirements pertaining to dechlorination and the discharge of chlorinated water.

Optimized distribution systems continuously review disinfectant residual values. Site investigations and reviews of historical data should be conducted when disinfectant residual concentrations fall below the minimum performance goals, are dissipating more quickly than expected, or where DBPs are occurring at high levels. Any unexpected changes are cause for investigation and possible action. It is important to note that actions affecting disinfectant residual and DBP concentration may be applied in the treatment plant as well as the distribution system. Adjustments are made to maintain disinfectant residual concentrations that meet optimization goals, while also minimizing DBP formation at all monitoring sites. Optimized systems consistently achieve the performance goals based on evaluation of data collected from optimized distribution system sampling locations. Although the majority of distribution system performance conditions are largely influenced by distribution system operations, do not overlook the potential impact of treatment plant operations on distribution system water quality. A discussion with treatment plant staff may help to elucidate the cause of some distribution system water quality issues. As stated in chapter 1, consider including a representative from the treatment plant staff on the distribution system self-assessment team.

Disinfectant Residual Optimization Goals

Disinfectant residual is the Partnership's key indicator of the water quality integrity of the distribution system. Utilities may assess the performance of the system with respect to disinfectant residual by comparing disinfectant residual and disinfection by-product sampling results with the Partnership's optimization goals for disinfectant residual and disinfection by-products. The following sections describe the Partnership's numerical goals for disinfectant residual and disinfection by-product performance and provide more information about the characteristics of optimized sampling locations throughout the distribution system.

Numeric Goals and Method of Calculation

Systems optimized with respect to disinfectant residual concentrations can demonstrate optimized performance by sampling at optimized disinfectant residual sampling locations and meeting the following disinfectant residual goals. The disinfectant residual goals were developed based, in part, on US Environmental Protection Agency regulatory requirements, with 4.0 mg/L being the maximum residual disinfectant level required by the US Environmental Protection Agency D/DBP rule and the minimum performance goals mirroring more conservative regulatory requirements applied in some regions to protect public health. Utilities are encouraged to review local regulatory requirements to ensure consistency with the Partnership's performance goals and should consult with the local regulatory agency in the case of inconsistency or concerns. In the case of local regulatory requirements for minimum disinfectant residual concentrations exceeding the Partnership's optimization goals, regulatory requirements supersede the Partnership's goals, which are listed below.

   The term *routine* is used to refer to disinfectant residual samples collected at an individual site on a regular schedule (continuously, daily, weekly, every two

Table 2-2.   Disinfectant residual performance goals

Parameter	Goal
Disinfectant residual, in the ranges specified, in 95% of monthly routine measurements.	Free chlorine: ≥ 0.20 mg/L and ≤ 4.0 mg/L Total chlorine: ≥ 0.50 mg/L and ≤ 4.0 mg/L Chlorine dioxide: ≥ 0.20 mg/L and ≤ 0.80 mg/L
Consecutive measurements	No consecutive residual measurements at individual routine sampling locations that are outside the disinfectant residual goals
Routine sampling locations (routine sampling locations are regularly scheduled and fixed locations that are selected to accurately represent distribution system water quality)	Sampling locations should include: System entry points (and booster station outputs) Problematic sites (dead ends, pressure boundaries) High water-age/low water-use sites Storage tanks (during water withdrawal) Additional sites are described throughout this section
Disinfection by-products (total trihalomethane (TTHM) and haloacetic acid (HAA5), as applicable)	Meet regulatory requirements for all samples tested Methods are implemented to maintain or reduce DBP concentrations, while maintaining adequate disinfectant residuals

weeks, or monthly). The intent of routine samples is for utilities to identify a targeted set of sample sites to be sampled as consistently as possible at similarly spaced intervals. Collectively, these sites should be representative of water quality conditions throughout the distribution system. It is also understood that utilities may need to vary sampling schedules occasionally because of system changes or other unforeseen reasons. Systems will rank the optimization status of disinfectant residual sampling locations and frequencies on a scale of 1-5 using the Partnership's data collection software. This will help utilities establish a baseline status for the optimization of sampling locations, so that future improvements in sampling (locations and frequencies) may be documented.

Utilities are encouraged to investigate sustained system changes between the regularly scheduled monitoring sampling. This might require collection of additional samples to evaluate consecutive low readings while improvements are being assessed. In this instance, only data collected on (or near) the predetermined regularly scheduled monitoring days should be used in the performance assessment. Utilities should document special study testing and remedial actions taken, and provide separate summaries for the self-assessment process.

Locations for Disinfectant Residual Testing

Disinfectant residual performance goals are based on data collected from the entire system over a 12-month period. The PSW philosophy is to use existing performance data, where possible, to complete the self-assessment. These data are collected, reviewed, and compared with the performance goals, both numerically and based on the optimization status of existing sample sites, so that an action plan for performance improvement may be created.

Distribution systems achieving full optimization expand on these existing data sets to reach a superior level of performance. The focus for these systems is shifted to more frequent testing at all system entry points (including master metering points from purchased supplies and treated wells), downstream of storage facilities, and other key optimization sites. The number of sample sites required for these systems depends on the size and configuration of the system. At a minimum, these utilities sample sites with a known high potential for improvement (described later), and should include other sites as needed for representation of all pressure zones.

Additional sampling sites may be required during changes in routine system operation. For example, systems that elect to perform a seasonal free chlorine "burn" to address nitrification issues should may need to assess the benefits of additional sample collection or the use of alternate sample collection sites during this period. Separate spreadsheets should be used for the submission of free chlorine and total chlorine data because of differences in the minimum concentration optimization goals for each parameter. Although a relatively common practice among utilities applying chloramine, particularly those in warmer climates, alternatives to the free chlorine "burn" exist that may provide utilities with additional options for preventing nitrification and/or reducing its impact, and that of the "burn," on distribution system water quality (AWWA 2013c).

For this performance assessment, systems must sample and test for disinfectant residual at representative sample sites including, at minimum:

- All entry points to the system daily (continuous chlorine monitors are used at fully optimized facilities). The daily median is used as the daily value from a continuous monitor
- Storage facilities (collected when water is withdrawn from the tank)
- Distribution system routinely monitored samples collected from optimized sampling locations (regularly scheduled: daily, weekly, or monthly)

If a distribution system does not currently sample all storage facilities or use a sampling plan that includes representative optimized sampling locations, the current monitoring approach should be evaluated during the self-assessment process to identify opportunities to improve sampling practices to help ensure delivery of the highest quality water.

Methods of Sample Collection and Measurement

Data entered for optimization evaluation purposes should reflect normal conditions and be a value measured *before* any nonroutine flushing or atypical tank operation. It is expected that disinfectant residuals will be representative of water "in the main" at the sample site. Collectively, and individually, the sample sites should be selected to represent water quality throughout the distribution system. Disinfectant residual sample taps will usually be flushed for about two to three minutes at a given site before determining the residual for a single sampling event, and sites with rather short service lines are ideal. Based on the data collected during this process, the utility will identify and make corrective changes that will consistently improve disinfectant residual levels at a given site (such as variations to storage operations or use of automatic flushing devices). It is expected that extra diagnostic testing may be performed at a given site between "official" monthly or weekly optimization testing for data entry ; however, excessive flushing or system changes should not be performed just before (or only because of) optimization data collection. Refer to the Appendix for an example disinfectant residual sampling procedure.

It is acceptable for distribution system sample sites to be added or removed over the course of the optimization data collection period. For example, if an action to add additional disinfectant residual sampling sites is identified during the self-assessment process, the system should not hesitate to add these sites during the data collection period. Additional information about changes in sample site locations may be provided in the self-assessment report, which also includes submission of a sampling site map. If there are blank fields in the data entry cells (in the Partnership's disinfectant residual data collection software) for a month, it will be obvious in the data entry sheet and summary table. Any gaps in sampling should be adequately documented, along with explanations provided for deleted or added sample sites and/or system changes (for example, site no longer accessible or addition of more useful sampling sites). All disinfectant residual measurements should be conducted with methods providing digital readings to two decimal places, where possible.

It is expected that samples for DBPs will be collected on a monthly or quarterly basis, with the maximum TTHM and HAA5 sample concentration recorded on any given day during the data reporting period. As with samples for disinfectant residual, utilities should be able to complete the annual data reporting process for DBPs using data currently available from samples collected by the utility. The utility is not required to collect additional DBP samples for the Partnership's data reporting or self-assessment process, unless this is identified as a specific action item by the utility.

Analytical Requirements

Disinfectant residuals will be determined using a colorimetric or amperometric measurement device with a digital readout (ANSI/AWWA C670-15). Color wheel measurements are *not* acceptable for this assessment. Data from continuous on-line disinfectant residual analyzers are ideal and should be included provided that they are frequently verified by field measurements. Data from these instruments must be representative and reported as the median measurement for the recording date. All disinfectant residual testing equipment must be properly calibrated, verified, and maintained. Calibration, verification, and maintenance activities should follow regulatory requirements or the manufacturer's recommendations, at a minimum, with regulatory requirements taking precedence. Calibration and verification documentation and records for each instrument should be maintained, current, and available. The data and the instruments must be verifiable and defensible.

It is suggested that instrument performance is verified periodically through the use of primary and secondary standards. Gel-type secondary standards are available for verifying the DPD chlorine chemistry. These standards may be used on a daily or weekly basis to verify instrument accuracy. Periodic use of a primary standard can be useful in verifying the entire analytical system, including the instrument, reagents, and analyst. Care must be taken to properly prepare a primary standard for chlorine residual because of the volatile and reactive nature of chlorine. This standard is best prepared in the laboratory. Consult *Standard Methods for the Examination of Water and Wastewater* (current edition) for more information regarding the preparation of primary standards for chlorine. Systems using online chlorine analyzers should be aware of any local regulatory requirements that may be associated with disposal of analyzer discharge, if applicable.

It is expected that samples for DBPs are collected and analyzed by a certified laboratory using an approved method. Colorimetric or screening methods for DBPs are not acceptable for optimization data collection reporting, although these methods may be useful for process control purposes. Online analyzers for DBPs may also provide useful and timely process control data for utilities, which may be used to supplement the data used for optimization data collection.

Performance Assessment Using Annual Report Data

Continuously delivering high-quality water to all users is the goal of an optimized distribution system. Disinfectant residual is the optimization parameter indicating water quality integrity for a distribution system. Disinfectant residual and DBP data are collected annually by the PSW from all participating water distribution systems. Statistics calculated from the test results are used by utilities to track their optimization progress and by the Partnership to quantify the program's long-term impact on distribution system water quality. These data are also a key component of the self-assessment process.

Data Requirements

The Partnership baseline and initial annual performance data include disinfectant residual and DBP results from routinely monitored distribution system sample sites. Disinfectant residuals (free chlorine, total chlorine, or chlorine dioxide) from throughout the distribution system are used.

Inputs for Baseline and Annual Disinfectant Residual Data Collection include the following:

- Median daily entry point disinfectant residuals (the average daily residual may be used but noted)

- Minimum daily residual value from all daily routine sample results

- The residual value and location (or location code) for *all* routine sample results not meeting the disinfectant residual goals

- The maximum daily total trihalomethane (TTHM) and sum of five halo-acetic acids (HAA5) sample results from all routine samples (on any day that samples are collected)

- The total number of disinfectant residual routine samples taken each day that sampling is performed

- The number of daily disinfectant residual routine sample results that fall outside the Partnership (or utility-specified) goals

- A determination of the utility's disinfectant residual sample site optimization status

The disinfectant residual data collection software automatically produces a summary table that includes the following and is displayed in Figure 2-1:

- Entry point—daily data entry used for monthly and yearly averages

- Distribution system—daily data used for monthly and yearly calculations and trends:
    - ▲ Total number of measurements—yearly and monthly
    - ▲ Number of measurements outside goals—yearly and monthly
    - ▲ Minimum disinfectant residual concentration—yearly and monthly (annual chart and frequency distribution chart)
    - ▲ Percent of daily minimum values not meeting goals—yearly and monthly
    - ▲ Number of sites with consecutive measurements not meeting goals—yearly and monthly
    - ▲ Exception grid—indicating sites where results do not meet goals by day
    - ▲ Exception summary—sites where results do not meet goals, yearly and monthly

- Distribution system—TTHM and HAA5 test results
    - ▲ Maximum values—yearly and monthly (annual chart)

	Annual	Jan	Feb	Mar	Apr	May	Jun	Jul	Aug	Sep	Oct	Nov	Dec
Entry Points Residual Average (mg/L)	0.45	0.45	0.45	0.45	0.45	0.45	0.45	0.45	0.45	0.45	0.45	0.45	0.45
Number of Routine Samples	729	61	56	62	60	62	60	62	62	60	62	60	62
Number of Routine Test Results Not meeting Goals	3	3	0	0	0	0	0	0	0	0	0	0	0
% Routine Test Results Not Meeting Goals	0.41	4.92	0.00	0.00	0.00	0.00	0.00	0.00	0.00	0.00	0.00	0.00	0.00
Minimum Daily Residual Value (mg/L)	0.02	0.02	0.40	0.40	0.40	0.40	0.40	0.40	0.40	0.40	0.40	0.40	0.40
Number of repeat non-conforming	1	1	0	0	0	0	0	0	0	0	0	0	0
TTHM Maximum (µg/L)	72.00	65.00			54.00			72.00			44.00		
HAA5 Maximum (µg/L)	20.00	20.00			15.00			12.00			20.00		

Figure 2-1. **Partnership for Safe Water example disinfectant residual summary table**

Refer to the Partnership's software guide for a detailed description of how data are entered into the current version of the disinfectant residual data collection software. The software, as well as the Partnership's annual data summary report, provides summaries of annual performance that may be helpful in determining trends and patterns in disinfectant residual and DBP performance. In addition to use of the Partnership's data collection software, utilities may wish to consider alternate means of analyzing disinfectant residual and DBP data, such as mapping data to visually identify potential areas of concern throughout the distribution system, as displayed in Figure 2-2.

Table 2-3 summarizes the PSW optimization goals for disinfectant residual and DBPs.

Optimized systems that meet all the numerical disinfectant residual performance goals should still perform the self-assessment to identify performance-limiting factors in the system, as there still may be opportunities for further optimization.

Status

To determine the performance status of the utility's distribution system with respect to disinfectant residual, develop and review the disinfectant residual summary tables, trend charts, and statistics for the most recent 1-year period. The evaluation of data from a more extended period will help utilities to better identify long-term trends in disinfectant residual and DBP performance. Also, develop and assemble data using the most recent version of the PSW disinfectant residual data collection software to assess performance. Based on this information, review the following items and provide responses to the following self-assessment questions:

- Do the disinfectant residual data meet the optimization performance goals? The goals are listed here and apply to both the entire distribution system and to each sampling location:

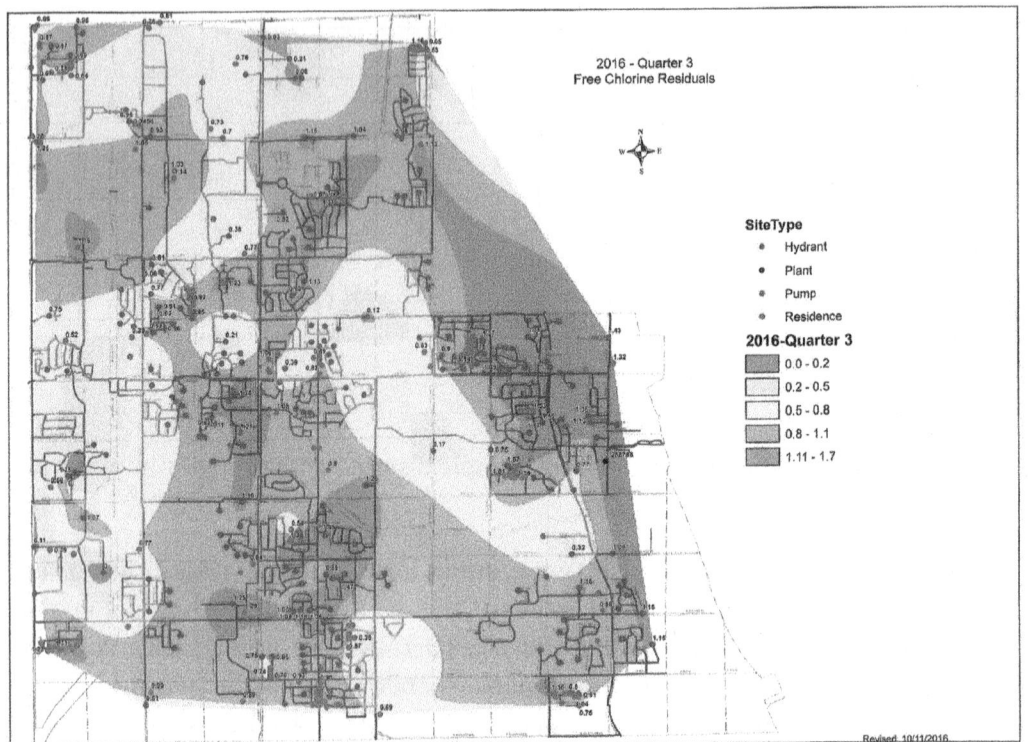

**Figure 2-2.   Example disinfectant residual map
(courtesy of Oak Creek Water & Sewer Utility)**

> ▲ Disinfectant residual in 95% of the monthly and yearly routine measurements
>
>> ▪ Free chlorine ≥ 0.20 mg/L and ≤ 4.0 mg/L
>>
>> ▪ Total chlorine ≥ 0.50 mg/L and ≤ 4.0 mg/L
>>
>> ▪ Chlorine dioxide ≥ 0.20 mg/L and ≤ 0.80 mg/L

- Are there any consecutive disinfectant residual measurements from optimized sample sites below the disinfectant residual goals?
- Does the utility have a system sampling map and schedule? If so:
  - ▲ Is utility staff sampling at the (previously described) optimized sampling locations?
  - ▲ Are sampling locations representative of distribution system water quality?
  - ▲ Are sampling locations tracked that repeatedly exhibit low disinfectant residuals?
  - ▲ Are nonroutine low-residual sample sites added to the future routine sampling schedule?

Table 2-3. Partnership for Safe Water optimization goals for disinfectant residual/ DBPs

Parameter	Goal
Summary	Systems meet all the numerical disinfectant residual performance goals using representative samples collected frequently from throughout the distribution system.
Disinfectant residual, in the ranges specified, in 95% of monthly routine measurements.	Free chlorine: ≥ 0.20 mg/L and ≤ 4.0 mg/L Total chlorine: ≥ 0.50 mg/L and ≤ 4.0 mg/L Chlorine dioxide: ≥ 0.20 mg/L and ≤ 0.80 mg/L
Consecutive measurements	No consecutive residual measurements at individual routine sampling locations that are outside the stated disinfectant residual goals
Sampling locations	Sampling locations must include all applicable optimization sampling locations, listed below: • All finished water entry points and metering points to other systems • Stage 1 and/or Stage 2 DBP sites • Initial Distribution System Evaluation (IDSE) or "IDSE-type" sites • Downstream of storage facilities (when water is being withdrawn) • Upstream and downstream of chlorine boosters • Areas served by older unlined cast iron mains • Low-flow areas • Known high water-age areas or other areas of concern
Disinfection by-products (TTHM and HAA5, as applicable)	Meet regulatory requirements for all samples tested Methods are implemented to maintain or reduce DBP concentrations, while maintaining adequate disinfectant residuals.

▲ Are performance improvement variables (chapter 3) used to reduce low residual recurrence?

• Is disinfectant residual testing performed using approved methods and digital testing equipment? Are values recorded to two decimal places, where possible?

• Are there online continuous chlorine analyzers in use throughout the distribution system? Are data from these analyzers collected and continuously displayed for operators by the supervisory control and data acquisition (SCADA) system?

Action

If performance-limiting factors are identified during assessment of the distribution system's performance assessment with respect to disinfectant residual, steps should be initiated to mitigate these concerns. This section offers some alternatives to consider for addressing any performance-limiting factors identified. Utility staff should consider these alternatives as the basis for beginning a comprehensive discussion of utility-specific action planning items.

- If the results of the performance assessment show the system consistently meets the disinfectant residual goals, then performance is considered optimized with respect to water quality. Even if water quality performance is optimized, the system should seek additional opportunities for operational and performance improvement by completing the remainder of the self-assessment process.

- If the utility meets the performance goals, an assessment should also be made to ensure that consistent performance can be maintained when the operators' process control capabilities are challenged. For example, new system facilities and stable water can lead to operator and administrator complacency resulting in performance degradation. (Complacency will be more fully assessed in chapter 6—Administration.)

- If disinfectant residual results are consistently lower than the Partnership's optimization goals, consider:

  ▲ Taking steps to reduce water age in the affected area of the system, such as automatic flushing and/or looping.

  ▲ Assessing the benefits of tank mixing or optimizing tank cycling if residual levels are consistently low in storage facilities or in areas under the direct influence of tanks.

  ▲ Applying booster chlorination, if appropriate.

  ▲ Verifying distribution system valve function to ensure valves are functional and operating as intended.

- If the system is unable to meet disinfectant residual performance goals also consider:

  ▲ Possible limitations with the process control program (see chapter 5).

    ▪ Consider developing or reviewing/updating Standard Operating Procedures (SOP) that provide operators with detailed instructions for actions to be taken in response to disinfectant residual sample results that do not meet the optimization goals or other internal standards.

  ▲ Possible system design limitations (see chapter 4).

    ▪ For example, do areas of low disinfectant residual correlate with other factors such as water age or pipe materials? Can an action plan be implemented to address these factors, such as increasing circulation in areas of high water age or implementing booster chlorination where appropriate?

▲ Applying or optimizing performance improvement variables (see chapter 3).

■ Examine the performance improvement variables covered in chapter 3 to determine if any of these areas may impact disinfectant residual performance. For example, nitrification can be a factor contributing to low total chlorine residual concentrations. If nitrification is a risk, the utility may wish to establish an action plan for sampling and response so that nitrification may be more quickly identified and addressed in the distribution system.

▲ Evaluating the level of administrative support (see chapter 6).

■ Limitations in administrative support can result in utilities having disinfectant residual data that are insufficient to enable determination of the system's optimization status. If this is the case, consider developing a plan that enables the utility to collect sufficient disinfectant residual data, such as through the installation of additional of online chlorine analyzers or more frequent collection of grab samples. Because of the potential for budgeting involved in this process, this may be a long-term action, best divided into smaller, shorter term goals.

Performance-Limiting Factor Summary

Table 2-4 summarizes factors related to disinfectant residual that may be limiting optimized performance. Indicate whether these factors are Optimized and Documented, Partially Optimized, or Not Optimized. Factors that are not fully optimized will be prioritized in chapter 7, Continuous Improvement Plan, and an action plan should be developed to address these areas.

PRESSURE MANAGEMENT

Maintenance of adequate pressure is an important barrier for protecting water quality in distribution systems (NRC 2006, Friedman et al. 2010). Although there are no federal regulations mandating specific distribution system pressures, most states have some requirements for minimum pressures. The "Ten State Standards" (Water Supply Committee of the Great Lakes—Upper Mississippi River Board of State and Provincial Public Health and Environmental Managers Recommended Standards for Water Works 2007) stipulate that water systems "shall be designed to maintain a minimum pressure of 20 psi (140 kPa) at ground level at all points in the distribution system under all conditions of flow." Additionally, these guidelines specify that the normal working pressure in the distribution system should be approximately 60 to 80 psi (410-550 kPa) and not less than 35 psi (240 kPa). However, specific pressure requirements may vary from system to system.

Pressure management is a key performance assessment criterion because the parameter is a critical barrier in maintaining the hydraulic integrity of the distribution system (NASTB 2006) and it influences other operational parameters. Utilities need to monitor, identify, investigate, and control pressure variations. In addition, systems should examine pipeline rehabilitation, replacement, and storage tank operational and maintenance practices, as these important factors also contribute to maintaining adequate pressure. Failure to consider all areas with the potential to impact pressure may result in limitations to optimizing pressure management.

Pressure management is influenced by main breaks and energy management. Pressure management may also influence customer complaints (those triggered by low pressure, erratic pressure, or large pressure fluctuations), water loss control, and backflow (and intrusion) susceptibility. Similarly, pressure management optimization can often be affected by flushing practices, particularly considering the impact of flushing on pressure stability, the potential reduction in pipeline carrying capacity caused by infrequent flushing, and the impact of valve and hydrant operation during flushing, which has the potential to create transient pressure events.

Understanding

Pressure is a common design factor used in positioning, selecting, and sizing pipes, tanks, pumps, and control valves. Regulatory agencies may have pressure standards or regulations that are used in planning, regulatory approval, and as operating guides. Utilities are advised to review regulatory requirements and to work closely with regulators as needed when assessing and optimizing pressure. Regulatory requirements must be met to participate in the Partnership. The PSW pressure optimization goals are almost always beyond what is required by regulations. If local or federal regulations are more stringent than the Partnership guidelines utilities must satisfy the regulations that apply and consider these regulations when completing the self-assessment process.

Through this self-assessment process, the utility collects and documents performance data for pressure management. Reviewing basic system configuration and operational information determines whether the system is capable of optimized (stable and acceptable) pressures throughout the system. The term *pressure management* describes the planning, monitoring, and operational measures that a utility uses to ensure distribution system pressures. Pressure variations

Table 2-4. Disinfectant residual assessment

Self-assessment category	Questions for gauging optimization status	Optimization status			Comments
		Optimized and documented	Partially optimized	Not optimized	
Disinfectant Residual	Do the disinfectant residual data meet the optimization performance goals? The goals are listed here and apply to both the entire distribution system and to each sampling location. Disinfectant residual in 95% of the monthly and yearly routine measurements: • Free chlorine: ≥ 0.20 mg/L and ≤ 4.0 mg/L • Total chlorine: ≥ 0.50 mg/L and ≤ 4.0 mg/L • Chlorine dioxide: ≥ 0.20 mg/L and ≤ 0.80 mg/L				Submit data collection spreadsheet.
	Are there any consecutive disinfectant residual measurements from optimized sample sites below the disinfectant residual goals?				
	Does the utility have a system sampling map and schedule? If so: • Is utility staff sampling at the (previously described optimized sampling locations)? • Are sampling locations representative of distribution system water quality? • Are sampling locations tracked that repeatedly exhibit low disinfectant residuals? • Are nonroutine low-residual sample sites added to the future routine sampling schedule? • Are performance improvement variables (chapter 3) used to reduce low residual recurrence?				
	Is disinfectant residual testing performed using approved methods and digital testing equipment? Are values recorded to two decimal places, where possible?				
	Are there on-line continuous chlorine analyzers in use throughout the distribution system? Are data from these analyzers collected and continuously displayed for operators by the supervisory control and data acquisition (SCADA) system?				

can suggest susceptibility to pressure transients that can lead to main breaks, increased leakage, and water quality problems due to backflow and intrusion. Pressure management optimization has the potential to result in cost savings through improved energy efficiency and prolonged infrastructure service life by reducing structural fatigue.

Utilities typically monitor pressure at storage tanks, pump stations, and valve chambers, but may not target areas most susceptible to pressure loss. It is important to consider that monitoring pressure at locations such as storage tanks is likely not sensitive enough to identify pressure events occurring elsewhere in the distribution system. Therefore, pressure trends from storage tanks can provide a false indication of stable pressure trends that may not be representative of actual distribution system pressure variations (LeChevallier 2014). Systems optimized for pressure management have continuous monitors at ideal locations throughout the distribution network, such as the high- and low-pressure sites within each pressure zone. Data from these continuous monitors should be evaluated in the self-assessment process. Optimized systems continuously evaluate pressure data to operate the distribution network. Utilities that do not have optimized monitoring systems should still complete the self-assessment using data from the pressure sensors it has, even if they are not ideally located. In this case, limitations in pressure monitoring should be identified as performance-limiting factors during the self-assessment, and an associated action plan should be created that outlines a feasible path to achieve optimized pressure monitoring. Most important, when additional pressure data become available from optimized locations, that data should be considered more representative. Therefore, it should

Table 2-5. References for disinfectant residual

Reference	Description
AWWA, 2012. *Standard Methods for Examination of Water and Wastewater.* 23rd ed. E.W. Rice, R.B. Baird, A.D. Eaton, L.S. Clesceri, editors. ISBN 978-0-87553-013-0.	Methods for chlorine analysis are described in detail with interferences and calibration-related issues.
AWWA, 2006. M20, *Water Chlorination and Chloramination Practices and Principles*, 2nd ed. Denver, Colo.: AWWA.	Manual focuses on the uses of chlorine and chloramines in municipal water treatment. Describes chemical properties, disinfection mechanisms, feed rates, handling, storage, safety, and DBP minimization techniques.
AWWA, 2013. M56, *Nitrification Prevention and Control in Drinking Water*, 2nd ed. Denver, Colo.: AWWA.	Manual provides extensive guidance and background useful for step and continuing oversight of chloramine operations. Nitrification causes, prevention, and remediation are described.
Black & Veatch Corporation, White, G.C., 2010. *White's Handbook of Chlorination and Alternative Disinfectants*, 5th ed. ISBN 9780470180983.	Extensive reference book explains principles, concepts, and applications of chlorine based disinfectants and alternatives.

be thoroughly evaluated to identify performance variations and possible additional action plans.

Components of an Optimized System for Pressure Management

Hydraulic models can be used to help determine the adequacy of the distribution system to maintain pressures within targeted ranges. Data describing the physical characteristics of the system, water demands, and system control status (initial and boundary conditions) are used to calibrate the hydraulic model. The model output provides the operational characteristics of the system, including pressures and flows. There are a number commercially available hydraulic modeling software packages, but all are based on the publicly available EPANET (USEPA 2008c). It is important that the model is properly calibrated.

Although most regulations require monitoring of distribution system pressure, there is little guidance on exactly how or where to conduct the testing. References for developing or conducting a pressure monitoring or pressure management program are limited. A number of studies, however, have focused on pressure monitoring and evaluation (Friedman et al. 2004, Gullick et al. 2005, Fleming et al. 2006). Studies indicate that both inadequate pressure and excessive pressures are generally due to a combination of problems (Deb et al. 2000), as summarized in Table 2-6.

Mains also lose their capacity because of leaks at joints, accumulations, corrosion holes, and tuberculation that reduces the carrying capacity of the pipe. There are many options, however, to manage and control distribution system pressure. For example, there are several guidelines and standards available for valve selection and installation, including ANSI/AWWA C512 Standards for Air Release, Air/Vacuum and Combination Air Valves for Waterworks Services (ANSI/AWWA C512-15); AWWA Manual M44 *Distribution Valves: Selection, Installation, Field Testing, and Maintenance* (AWWA 2016c); and AWWA Manual M51 *Air-Release, Air/Vacuum, and Combination Air Valves* (AWWA 2016d).

It is important that valves be appropriately designed and installed because failure to do so can produce secondary pressure surges. In addition, a number of reports on pressure management devices, together with rationale of choosing appropriate strategies, can be found in standard engineering texts on pump

Table 2-6. Causes of excessive and inadequate distribution system pressure

Causes of excessive pressure	Causes of inadequate pressure
Lack of pressure-reducing valves	Changes in valve settings
Changes in valve settings	Decreased pump heads
Increased pump heads	Reduced tank levels
Increased tank levels	Increased water demand
Inadequately sized mains	Oversized mains
	Main breaks or large leaks

design, water distribution system pipeline design, and pressure transient analysis (Karassik et al. 1976, Larock et al. 2000, Thorley 2004). Additional information about hydrants, valves, and pumping facilities may be accessed in chapter 4.

The benefits of pressure management can more than offset the costs associated with implementing sound pressure management strategies. Management of distribution system pressures can reduce main breaks (and collateral damage and costs), system leakage, and energy consumption. Management to assure a safe minimum pressure is consistently provided throughout the distribution system can reduce the risks of contamination from intrusion or backflow, increasing public health protection and potentially saving the significant costs associated with a waterborne disease outbreak originating from low distribution system pressures.

Pressure Management Goals

Objective performance indicators are used to track the degree to which pressure is managed in distribution systems. These are primarily adapted from Ten State Standards (GLUMRB 2007) and AWWA Manual M32 (AWWA 2012b) pressure design criteria. There are several core performance indicators used to assess whether distribution system pressures are managed properly, including:

- Normal Water Demand Periods: 35 psi average minimum (monthly average of daily minimums)
- Peak Water Demand and Fire Flow Periods: 20 psi minimum (hourly minimum during simultaneous peak water demand and fire flow conditions)
- Emergency Conditions: 0 psi minimum (during emergencies such as main breaks and power outages), response plan to prevent the possible spread of contamination, limit frequency, duration, and extent of low pressure events

Because it is not feasible to monitor pressure at every location in the distribution system, the optimization criteria are applied at two critical locations in each pressure zone: locations most representative of high and low pressure. Optimization criteria are also applied to existing water level transmitters and pressure monitors that might be at nonideal locations such as storage tanks, pump stations, and valve chambers (less representative and sensitive locations, but still valuable). When reviewing data from optimized and nonoptimized locations, the self-assessment team should place more weight on the validity of the data generated from pressure sensors sited in optimized locations.

Numerical Pressure Goals and Method of Calculation

Pressure management optimization goals take into account the widely accepted minimum of 20 psi, as well as utility-specific goals. The numerical optimization goals for pressure are listed in Table 2-7.

Table 2-7. Optimization goals for pressure management

Parameter	Goals
Minimum Pressure	≥ 20 psi in 99.5% of measurements (daily minimums from pressure readings)
Maximum Pressure	Does not exceed utility specified maximum in 95% of measurements
Pressure Fluctuations	Do not exceed utility specified maximum pressure fluctuation range in 95% of measurements

Pressure must be monitored continuously. The utility should strive to maintain a minimum pressure at or above 20 psi at all times. For the purpose of this performance assessment, this goal is measured by recording the daily (from hourly averages or instantaneous individual pressure measurements) minimum pressure from any continuous pressure measurement site (the site location is noted). Systems optimized with respect to pressure management maintain a daily minimum pressure at or above 20 psi in 99.5% of these measurements (daily *system* minimum may be below 20 psi for 2 days in a year).

After careful review of system piping and recommended operating pressure, the system should establish a utility-specific maximum pressure goal and maximum pressure fluctuation range goal. These goals should be carefully assessed and documented, and their importance recognized by all members of the self-assessment team. Once established and accepted by staff throughout the organization, the utility will then be able to monitor ongoing performance to facilitate optimization.

The performance assessment calculates pressure statistics using the minimum and maximum daily values from pressure sensors located throughout the distribution system (ideally located at representative low and high pressure sites), as well as the maximum single-site pressure fluctuation range. If pressure sensors are not located at ideal sites, then use the lowest and highest pressure sites *available*. If nonideal data are used, the system is encouraged to implement actions to improve this situation by identifying this as a performance-limiting factor and including appropriate steps for its continuous improvement in an associated action plan (chapter 7). This is necessary even if the existing data from nonideal locations indicate acceptable pressure trends.

Limiting pressure variations (pressure range) may be the most challenging element of optimizing this criterion. Although transient (very short term) pressure fluctuations may be important, it is not the intent to use these values in the performance assessment. Pressure minimums and maximums of at least 5-15 minutes duration should be used as a general guide for consideration in the self-assessment process. As a general rule, if the pressure fluctuation is of a sufficient duration to which the utility staff would take action to respond, it is recommended to consider evaluating the event as part of the performance assessment process. It is not intended that extremely short term pressure transients be considered in the self-assessment process, although it is important for utility staff to be aware of these events, if they occur, and their potential causes. Some systems may not be able to achieve the optimization goals for pressure management prior to submitting a Phase III self-assessment completion report. These systems would not be fully optimized as defined by the PSW, but would still be considered for the Phase III Directors Award status, as long as this performance-limiting factor is recognized and appropriate action plans for future improvement are developed and documented within the self-assessment report.

Monitoring Locations

Pressure monitoring, ideally, is conducted at two key locations in each pressure zone. The locations are selected to be representative of high and low pressure sites in each zone. Typically, pump stations and elevated storage tanks are locations at the highest and lowest elevation within a pressure zone ; however, these sites could be influenced by other factors such a pipe diameter, flow patterns, and valve operations (note: for very small pressure zones with little change in elevation, a single monitoring location can be used). Using a calibrated hydraulic model can help identify the best monitoring locations. Data from key locations supplements routine pressure monitoring from other monitors that are used for operational control (such as storage tanks, pump stations, and valve chambers).

Method of Measurement

Permanently installed pressure sensors are used, at key locations, to continuously monitor and record the pressure data. Data from such installations are continuously captured by computer (SCADA) systems, and operators are alerted when conditions fall outside normal operating ranges. Because some distribution systems do not have pressure sensors permanently installed in the ideal locations, it is acceptable to conduct the performance assessment using data from nonideal locations or portable pressure sensors when necessary. However, as previously stated, this should be identified as a performance-limiting factor and addressed in the action plans for future improvement. Systems without *any* pressure sensors cannot complete the Partnership's self-assessment process.

Analytical Requirements

All pressure monitoring instrumentation should be appropriately scaled for the application, and appropriately installed, calibrated, and maintained as per manufacturer's recommendations or better. Instrumentation found to be out of calibration during the annual calibration should be checked again within six months to assure reliability. Calibration should adhere to procedures recommended by the manufacturer. Documentation for instrument calibration and verification records should be maintained, current, and available. It may be helpful to maintain a minimum level of onsite inventory for critical spare parts for analytical instrumentation, to help prevent downtime for sensors providing critical pressure information.

Pressure Management Performance Assessment

Pressure management is the Partnership's key indicator of the hydraulic integrity of the distribution system. Utilities may assess the performance of the system with respect to pressure by comparing distribution system pressure measurements with the Partnership's optimization goals for pressure. The following sections describe the Partnership's numerical goals for pressure performance and provide more information about optimized sampling and data collection practices.

Data Collection

Pressure is a key system control measure used by system operators. Values from locations throughout the system are continuously monitored and evaluated. Changes in pressure require immediate response since the cause may indicate an emergency situation, which is why real-time pressure monitoring is critically important to distribution system operation. The performance assessment uses pressure minimum and maximum values to evaluate performance from continuous pressure monitoring data.

Pressure data are collected from at least two sites (low and high pressure) in each pressure zone. Discrete values from sensors' continuous data should be carefully examined to exclude values that are not representative of actual pressure in the system, and documentation of the reasons for certain data exclusion is necessary. Data that are the result of power outages, produced while instruments are being calibrated, and other identifiable causes should not be used in the performance assessment. Data fluctuations of short duration but representative of actual system pressure should not be excluded. These values may be providing important information about distribution system pressure transients.

Systems supplement pressure monitoring from the two critical locations in each pressure zone (most representative) with routine pressure monitors that are installed at less representative locations, such as storage tanks, pump stations, and valve chambers. Criteria for low (less than 20 psi) pressure must be achieved at *all* monitoring locations. To complete the Phase III self-assessment process, data from nonideal sensor locations may be used, if necessary. This performance-limiting factor should be identified in the self-assessment completion report and addressed as utilities move forward in pursuit of achieving optimized distribution system performance.

Assessment Software

The performance assessment for pressure management spans a one-year period of operation, and one year of pressure monitoring data is the minimum required for submission of the self-assessment completion report. The year can begin in any month but must include a complete year (12 months) of operation. If there are gaps in data from technical malfunctions, these must be explained. Any gaps longer than a few days may require the assessment to be repeated so that continuous operation is reflected by the data.

The pressure assessment data should be compared with a calibrated distribution system hydraulic model, when available. Ideally, pressure data should be compared with the hydraulic model on an annual basis (under identical conditions with extended period simulation of the hydraulic model) with a goal of agreement between measured and modeled data of 5 psi or less. Additionally, periodically evaluate the potential for surge pressures at each pump station or when any significant changes are made to a pump station (for example, larger pumps or motors installed or additional pumps added).

Daily maximum and minimum pressure values should be entered into the Partnership's pressure data collection software, along with the maximum site pressure fluctuation range, the number of sensors used by the system, and the number of sensors measuring pressures of less than 20 psi. Data from both optimized sampling locations and nonideal sampling locations may be included in the data entry process. Utilities identify the location of pressure sampling sites by code within the software program. For security purposes, the specific locations are not required to be identified when the data are submitted to the PSW.

Although all of the distribution system pressure data may be combined on a single spreadsheet, utilities with multiple pressure zones are encouraged to submit separate data collection spreadsheets for each pressure zone. This is particularly important for distribution systems that may have significant variations in the operational parameters for pressure management between pressure zones.

Data Analysis

The Partnership's pressure data collection software calculates and displays statistics for evaluation and comparison with the performance criteria. The pressure data collection software requires the following data inputs:

- Minimum daily pressure value (from permanent pressure sensors) from the distribution system low pressure site or each pressure zone
- Maximum daily pressure value (from permanent pressure sensors) from the distribution system high pressure site or each pressure zone
- Maximum single-site pressure fluctuation range
- Number of sensors in use in the system

- Number of pressure sensors reading outside of the Partnership's or utility-determined optimization goals
- Identification of the minimum and maximum pressure sites
- Pressure monitoring site optimization ranking

A summary of statistics and trend charts are generated for the data entered. As previously stated, it is recommended that systems with multiple pressure zones create separate data collection spreadsheets for each pressure zone, particularly if the zones vary significantly in their pressure management objectives. The purpose of these data outputs is to aid the utility in the assessment of its distribution system optimization status with respect to pressure. These data are also provided to the PSW for evaluation of overall program performance with respect to pressure management.

The software calculates and displays the following information:

- Monthly and annual average minimum pressure, maximum pressure, and maximum pressure fluctuation range
- Percent of minimum measurements that are < 20 psi (monthly and annually)
- Percent of maximum measurements that are greater than the utility-specified goal (monthly and annually)
- Percent of single site pressure fluctuations that are greater than the utility-specified goals (monthly and annually)
- Summary of sites with multiple daily minimum values < 20 psi
- Daily minimum value graph
- Frequency distribution of daily minimum pressure values

An example of the summary table generated by the Partnership's pressure data collection software is displayed in Figure 2-3.

	Annual	Jan	Feb	Mar	Apr	May	Jun	Jul	Aug	Sep	Oct	Nov	Dec
% Daily Minimum < 20psi	0.5	6.5	0.0	0.0	0.0	0.0	0.0	0.0	0.0	0.0	0.0	0.0	0.0
% Daily Maximum > goal	0.0	0.0	0.0	0.0	0.0	0.0	0.0	0.0	0.0	0.0	0.0	0.0	0.0
% Daily Single Site Range > goal	0.0	0.0	0.0	0.0	0.0	0.0	0.0	0.0	0.0	0.0	0.0	0.0	0.0
Average Daily System Min.	21.9	20.8	22.0	22.0	22.0	22.0	22.0	22.0	22.0	22.0	22.0	22.0	22.0
Average Daily System Max.	45.0	45.0	45.0	45.0	45.0	45.0	45.0	45.0	45.0	45.0	45.0	45.0	45.0
Average Daily System Range (Max)	23.0	23.0	23.0	23.0	23.0	23.0	23.0	23.0	23.0	23.0	23.0	23.0	23.0
Avg. Number Pressure Sensor Locations in Service	2.0	2.1	2.0	2.0	2.0	2.0	2.0	2.0	2.0	2.0	2.0	2.0	2.0
Number of Pressure sensors below 20	2.0	2.0	0.0	0.0	0.0	0.0	0.0	0.0	0.0	0.0	0.0	0.0	0.0
Number of Pressure Sensor locations Repeatedly below 20 psi	0	0	0	0	0	0	0	0	0	0	0	0	0

Figure 2-3. Example pressure data collection software summary table

It is possible that low- or negative-pressure events may occur within a distribution system under emergency conditions, particularly if a main was depressurized during a repair or a major power outage. In these unusual situations, appropriate precautions must be taken to protect public health, such as issuing a boil water advisory or disinfecting and flushing repaired pipelines. The self-assessment should include a discussion of the policies and procedures used to avoid and respond to a low- or negative-pressure event, with supporting information as appropriate.

Data Assessment

Figure 2-3 provides an example of the summary table created by the Partnership's pressure data collection software. Although the software may be applied for use across the entire distribution system, it is recommended that data for each pressure zone be collected in a separate software file to generate a unique statistical data summary for each zone. These data can be used both to evaluate the optimization status, with respect to pressure, for the entire distribution system, as well as each pressure zone. Fully optimized systems assess pressure management for each zone using the method described here for the entire system. System pressure management status is assessed by comparing calculated values with the Partnership's optimization goals. A fully optimized distribution system would meet the optimization goals for the entire system, as well as for each individual pressure zone.

The performance assessment software also creates a chart of daily minimum pressure and a frequency distribution of daily minimum pressure measurements. These charts can be beneficial in helping utility staff to quickly evaluate pressure management performance over a 12-month period. Graphical display of the data makes pressure values below 20 psi readily evident and easy to identify. Using the minimum pressure frequency distribution chart also makes it easy to identify whether more than 0.5% of minimum pressure values are less than 20 psi. The PSW calculates similar statistics, in aggregate, for all participating utilities, so that they may be used for comparison with individual distribution system performance. Utilities are encouraged to use the trend charts and summary tables generated by the Partnership's data collection software to compare overall performance from year to year.

Table 2-8 summarizes the PSW optimization goals for pressure management. Systems determined to be optimized, with respect to pressure management, are encouraged to complete the assessment to identify system performance-limiting factors, as there still may be opportunities for further improvement. For example, systems with tight control over variations in system pressures will still need to examine other areas of distribution system operation, such as storage tank turnover, because poor mixing and turnover in storage tanks (caused by keeping storage facilities "full") can potentially result in loss of disinfectant residuals and water quality issues, including nitrification. Systems

Table 2-8. Optimization goals for pressure management

Parameter	Goals
Summary: Systems meet all of the numeric performance goals for each pressure zone and for the overall system	
Minimum Pressure (under normal operating conditions, including maximum hourly demand and fire flow)	≥ 20 psi in 99.5% of measurements (daily minimums from pressure readings)
Maximum Pressure	Does not exceed utility specified maximum in 95% of measurements
Pressure Fluctuations	Do not exceed utility specified maximum pressure fluctuation range in 95% of measurements
Monitoring	Pressures are continuously monitored by permanent sensors located at the low and high pressure sites (confirmed by studies) within each zone. Additional sensors may be employed for operational purposes. Pressure values at all sites must meet the performance assessment goals. Pressure sensors are routinely calibrated and maintained according to manufacturer recommendations.
Standard Operating Procedures	The system has standard operating procedures for maximum system pressures, pressure fluctuations and infrequent emergencies that result in localized pressures not meeting the goals. These procedures must protect the public health with the endorsement of the regulatory authority responsible for the system. The system must demonstrate use of these procedures when necessary.

with tight pressure control may also consider reducing excessive system pressures (while still meeting all of the optimization goals) to reduce water loss and minimize main breaks.

Status

To determine the status of the distribution system with respect to pressure management, collect the appropriate pressure data, enter the data into the Partnership's pressure data collection software, and review the results generated. Based on this information, in combination with distribution system operational practices, review the following items:

- Does the system meet all of the pressure management optimization goals?
  - ▲ 20 psi minimum (under normal operating conditions including maximum hourly demand and fire flow) in 99.5% of the daily minimum measurements.

    ▲ Maximum pressure does not exceed utility specified maximum in 95% of measurements

    ▲ Pressure fluctuations do not exceed utility specified maximum pressure fluctuation range in 95% of measurements

- Is pressure monitored at a minimum of two critical sites in each pressure zone (areas of high and low pressure)? Is the pressure at these sites monitored continuously? Are the instruments properly installed and routinely calibrated?

- Does pressure monitoring include maximum day demand flow, fire flow events, and emergency situations (such as a main break or power outage)?

- If online pressure monitoring is available, is the data analysis configured to alarm the operator when low-pressure (<35 and/or 20 psi) or high-pressure spikes occur?

- Are pressure fluctuations monitored and investigated, and are procedures used to reduce the range and duration of pressure variations?

- Are there preapproved procedures (written SOPs) that help prevent low-pressure events and to protect public health in case of pipeline depressurization (from main breaks or due to power outages)?

- Is the operator aware of conditions that can cause low or high system pressures? Are SOPs available that account for routine and nonroutine operations that might affect pressure? Are maximum system pressures and pressure fluctuations documented?

Action

If problems are indicated during the status review of the distribution system's pressure performance assessment, steps should be initiated to mitigate these problems. This section offers some alternatives to consider for addressing identified performance-limiting factors. Utility staff is not limited to these considerations and are encouraged to assess the benefits of additional potential actions for realizing pressure performance improvements in the distribution system.

- If adequate and accurate pressure data was used to complete the pressure performance assessment, and the results show that the system consistently meets the pressure management goals, then the pressure performance may be considered excellent for the particular period evaluated. Any activities identified in the remaining chapters of the self-assessment that impact pressure management will reflect further refinement of operation or administration practices rather than result in significant pressure control improvements.

- If the amount of pressure data is insufficient to make a determination of the system's optimization status, or placement of pressure sensors is at nonideal locations, an action plan should be developed to address pressure sensor placement and data collection. The action plan should include both short and long term actions and should include budgeting for pressure monitoring instrumentation, if required.

- Even if the utility meets the performance goals, an assessment should also be made to ensure that the consistent performance is not related solely to a relatively new system that rarely challenges the operators' process control capabilities. New system facilities can lead to operator and administrator complacency leading to performance degradation.

- If the system is unable to meet pressure performance goals, consider addressing:

  - ▲ Possible limitations with the process control program (see chapter 5).

  - ▲ Possible limitations in system design (see chapter 4).

  - ▲ Applying or optimizing performance improvement factors that have the potential to affect pressure (see chapter 3).

  - ▲ Evaluating the level of administrative support for optimizing pressure management activities (see chapter 6).

Performance-Limiting Factors Summary

Table 2-9 summarizes factors related to pressure management that may be limiting optimized distribution system performance. Indicate whether these factors are Optimized and Documented, Partially Optimized, or Not Optimized. Factors that are not fully optimized will be prioritized in chapter 7, Continuous Improvement Plan, and an action plan should be developed to address these areas.

Table 2-9.   Pressure management assessment

Self-assessment category	Questions for gauging optimization status	Optimization status			
		Optimized and documented	Partially optimized	Not optimized	Comments
Pressure Management	Does the system meet all of the pressure management optimization goals? • 20 psi minimum (under normal operating conditions including maximum hourly demand and fire flow) in 99.5% of the daily minimum measurements • Maximum pressure does not exceed utility specified maximum in 95% of measurements • Pressure fluctuations do not exceed utility specified maximum pressure fluctuation range in 95% of measurements				Submit pressure data collection spreadsheet. Report utility specified goals for pressure maximum and range.
	Is pressure monitored at a minimum of two critical sites in each pressure zone (areas of high and low pressure)? Is the pressure at these sites monitored continuously? Are the instruments properly installed and routinely calibrated?				
	Does pressure monitoring include maximum day demand flow, fire flow events, and emergency situations (such as a main break or power outage)?				
	If online pressure monitoring is available, is the data analysis configured to alarm the operator when low pressure (<35 and/or 20 psi) or high pressure spikes occur?				
	Are pressure fluctuations monitored and investigated, and are procedures used to reduce the range and duration of pressure variations?				
	Are there preapproved procedures (written SOPs) that help prevent low-pressure events and to protect public health in case of pipeline depressurization (from main breaks or due to power outages)?				Submit negative pressure response plan.
	Is the operator aware of conditions that can cause low or high system pressures? Are SOPs available that account for routine and nonroutine operations that might affect pressure? Are maximum system pressures and pressure fluctuations documented?				

References

Table 2-10. References for pressure management

Reference	Description
National Research Council, 2006. Drinking Water Distribution Systems: Assessing and Reducing Risks. Washington, D.C.: National Academy Press.	Describes physical, hydraulic, and water quality integrity for distribution systems.
USEPA, 2008. EPANET.	Publicly available hydraulic modeling software package.
Gullick, R.W.; LeChevallier, M.W.; Case, J.; Wood, D.J.; Funk, J.E.; & Friedman, M.J., 2005. Application of Pressure Monitoring and Modeling to Detect and Minimize Low Pressure Events in Distribution Systems. *JWSRT—AQUA* 54(2):65-81.	Extensive pressure monitoring of a distribution system using seven electronic data loggers for 1.4 years found nine low pressure events (< 20 psi). The model simulations showed that installing hydro-pneumatic tanks eliminated low pressure threats.
Friedman, M.; Radder, L.; Harrison, S.; Howie, D.; Britton, M.; Boyd, G.; Wang, H.; Gullick, R.; LeChevallier, M.; Wood, D.; & Funk, J., 2004. AwwaRF #2686. Verification and Control of Low Pressure Transients in Distribution Systems.	Conducted long-term pressure monitoring at three locations in seven distribution systems. Short term monitoring focused on pump tests, high demand situations, valve operations, and surge tank operations.
Great Lakes Upper Mississippi River Board of State and Provincial Public Health and Environmental Managers (GLUMRB), 2007. Recommended Standards for Water Works. Albany, NY: Health Research Inc., Health Education Services Division.	Section 8.2.1 outlines the requirements for system design for pressure management.
Fleming, K.K.; Gullick, R.W.; Dugandzic, J.P.; & LeChevallier, M.W., 2006. AwwaRF #3008. Susceptibility of Distribution Systems to Negative Pressure Transients.	Examines characteristics of distribution systems that increase the susceptibility to negative pressure transients.
Karassik, I.J.; Krutzsch, W.C.; Fraser, W.H.; & Messina, J.P. (editors.), 1976. *Pump Handbook*. New York, N.Y.: McGraw-Hill Education.	Textbook on pump selection, sizing, and operations to manage pressures and avoid pressure fluctuations.
Larock, B.E.; Jeppson, R.W.; & Watters, G.Z., 2000. *Hydraulics of Pipeline Systems*. New York, N.Y.: CRC Press.	Textbook on pipeline hydraulics. Outlines pressure management options for sizing of pipelines.
LeChevallier, M.W., 2014. WRF #4321. Pressure Management: Industry Practices and Monitoring Procedures. Denver, Colo.: Water Research Foundation.	
Thorley, A.R.D., 2004. *Fluid Transients in Pipeline Systems,* 2nd ed. Herts, England: D. & L. George, Ltd.	Textbook on managing fluid transients that can result in variations in pressure.

Reference	Description
Deb, A.K.; Momberger, K.A.; Hasit, Y.J.; & Grablutz. F.M., 2000. *Guidance Management of Distribution System Operations and Maintenance.* Denver, Colo.: AWWA and AwwaRF.	Outlines operations and maintenance procedures for a variety of distribution system issues including pressure management.
Walski, T.M.; Chase, D.V.; Savic, D.A.; Grayman, W.M.; Bechwith, S., & Koelle, E., 2003. *Advanced Water Distribution Modeling and Management.* Watertown, Ct.: Haestad Methods Inc..	Textbook on hydraulic modeling used to evaluate pressure management options. Outlines model calibration procedures and requirements.

MAIN BREAK FREQUENCY

Main break frequency (Main Breaks and Leaks) has been selected as the performance criterion for demonstrating optimization for the physical integrity of the distribution system infrastructure. The content provided in this section of the Guide is largely drawn from the Water Research Foundation project #4109 final report (Friedman et al. 2010). To complete this portion of the self-assessment, utilities analyze assembled leak and break data to assess performance and determine their status with respect to an optimized distribution system physical infrastructure. The objective of this part of the assessment is to quantify breaks and leaks and compare current performance with historical performance, as well as the Partnership's optimization goals. Utilities will also attempt to correlate breaks and leaks with other distribution system variables and assess their main break response procedures.

Main break frequency, in this assessment, refers to both breaks and leaks as defined in this portion of the chapter. Note that the data refer to reported breaks and leaks and do not include those identified as part of a utility's active leak detection program. A robust leak detection program is an important element of optimized distribution system operations and is addressed in greater detail in the Water Loss Control section of chapter 3.

Understanding

Main break frequency is a primary indicator of distribution system infrastructure condition, because a reduction in the occurrence of main breaks is a critical element in maintaining distribution system integrity. Main break events have a large and visible impact on several other key operational parameters and can also have an impact on public health protection. For example, main breaks can adversely affect adequate pressure maintenance, whereas poor pressure control can, conversely, cause an increased occurrence of leaks and breaks. Hydrant and valve maintenance, pipeline rehabilitation and replacement, storage tank maintenance and operation, and internal and external corrosion control are also important factors that can contribute to managing main breaks. Consideration of multiple distribution system variables has the potential to result improved control and a reduction in the potential impact of main breaks.

Pressure management and water loss control are also related to main breaks. Effective water loss control programs, such as leak detection and water audit programs, can help utilities identify hidden leaks and facilitate early repair before costs and consequences increase. Limitations in main break management can also impact water loss. Customer complaints and perceptions about utility performance may be influenced by the frequency, duration, and extent of reported main breaks. Main breaks themselves, or poor response to main break events, can lead directly to system contamination at the break site or indirect contamination by depressurizing the system and causing backflow at other locations. Therefore, the most significant impacts of reported main breaks are usually normal service disruption, increased customer complaints, and the risk of contamination, rather than water loss.

Reported main breaks are events that disrupt normal operations or cause complaints from customers or the community, requiring immediate attention. However, many utilities may not currently evaluate this critical indicator in greater depth, representing a missed opportunity to improve and optimize distribution system operation and maintenance. Optimization of main break frequency can result in improved customer satisfaction, reduced costs for emergency repairs, reduced levels of water loss, stabilized pressures, and improved water quality.

Utilities typically monitor reported main breaks and their associated direct costs, but may not always document and analyze the indirect costs and root causes of main breaks before planning for improvements. The term *main breaks management* is used in this self-assessment to describe the planning, monitoring, and operational measures that a utility uses to identify, document, and respond to leaks and breaks within the distribution system. The following section describes the components of a system that is optimized with respect to main breaks management.

Components of an Optimized System for Main Breaks Management

Comprehensive Tracking of Main Breaks and Leaks

Although there are no uniformly established industry standards for data collection, many utilities have developed their own data collection forms for breaks and leaks. The most effective tracking systems are those that have a standard form with defined data fields that are fully understood and consistently applied by both field and office personnel. Forms may be hardcopy or electronic in nature—the most important factor is that they are used consistently and accurately by utility staff to capture and record relevant information about main breaks. Two reports from Deb (1995, 2002) provide sample forms for data collection and descriptions of the data elements that should be collected.

Deb (2002) classifies main break data as field, office, testing, and cost data. Field data are the elements that can be observed and recorded while investigating or repairing the leak or break. Examples of field data include:

- Location of break
- Size of break and pipe involved
- Pipe material
- Visible condition upon inspection
- Failure type

Office data are information usually available in company records or from sources other than the field responders. Examples of office data include:

- Year pipe was installed (age of pipe)
- Typical pressure and flow conditions in area
- Weather data

Testing data includes the results of any tests that may have been performed by staff or an outside laboratory that are associated with a break event. These data may include:

- Soils testing
- Materials testing

Main break data must be collected on-site during the investigation. Some data, such as asset age, may need to be added to the reported break record several days or weeks after the event. When active leakage control activities identify an unreported leak, the same standard data collection form should be initiated. This event should be coded so it is not included in the performance goal calculation of "reported" leaks and breaks.

Some techniques that utilities use for detecting main breaks include permanently installed, continuous acoustic monitors at locations in their system to identify unreported leaks and breaks. Data from such installations would be continuously collected and analyzed by software systems, and field personnel would be alerted when a suspected leak has been identified. Also, use of district metered areas (DMAs) with supervisory control and data acquisition (SCADA) systems would identify when conditions fall outside normal operating parameters and prompt utility staff to begin leak investigation. Other more traditional methods such as periodic leak detection surveys or temporary installation of acoustic monitors and flow meters may also be used to identify unreported leaks. Water utilities should determine the combination of methods that is most economically and technically appropriate to meet the utility's individual leak reduction goals.

Cost data are also added after the event when labor, material, and contractor costs are processed and after indirect costs, such as damage claims and customer service interruptions, are evaluated and determined. When collecting data associated with a main break, consider the benefits of collecting samples that are only available during the infrequent times when the pipe and the surrounding soils are exposed. Even if they are not immediately required at the time of the break, it may be beneficial to consider proactively collecting these soil and materials samples, while they are available. Refer to the External Corrosion Control section of chapter 3 for additional information.

Analysis of Leak and Break Data

The analysis of main break and leak data is a critical element in managing water distribution system infrastructure. Deb (2002) describes five categories of analysis to consider:

1. Material (pipe type, diameter, joint type)
2. Spatial (location)
3. Temporal (by season, trend over years, correlation with weather and soil or water temperature)
4. Economic (cost of repair vs. replacement)
5. Mechanistic (evaluation of corrosion rates, graphitization, applied loads)

For all of these analyses, the objective is to identify mains with high average failure rates. Usually, these leaks or breaks will have similar failure characteristics. The utility then investigates and evaluates the causes of failure and selects the most appropriate actions to implement to reduce the rate of failure. Sometimes, there may be operational changes, such as surge control or pressure reduction, that can reduce the incidence of failures. In other instances, the only viable action to reduce recurrent failures is scheduled replacement or rehabilitation of the affected pipeline section.

Main Break Response

Effective response to main breaks requires performance of many related tasks, often in quick succession and involving several employees. Optimized utilities have written operating procedures, which are regularly reviewed and updated, and provide training to their employees for investigating, locating, isolating, and repairing leaks and breaks. As part of these procedures, other utilities, such as gas or electric, must be notified and their facilities marked before excavation begins. In addition, employees must be able to work safely, document the leak or break data, properly disinfect and protect the main from contamination, and restore service.

Water Research Foundation Report #4307 (Effective Microbial Control Strategies for Main Breaks and Depressurization, Kirmeyer et al 2014) categorizes main breaks into four categories, based on the severity of the event and the level of risk each presents to microbial contamination. For example, addressing a Type 1 break (the least severe category in which positive pressure is maintained at all times) is typically accomplished by excavating a pit to below the break, repairing the pipe under pressure with disinfected parts, and measuring disinfectant residual levels upon completion. No additional bacteriological sampling is required, nor is issuance of a Boil Water Advisory. Conversely, a Type 4 break (the most severe category, in which there is widespread depressurization) may require issuance of a boil water advisory, flushing, extensive disinfection and verification, and some level of crisis communication and response. A full description of appropriate actions to address each type of main break may be accessed in the report. It is important that utilities establish SOPs for main break response and follow the appropriate procedures for addressing main breaks in the distribution system.

If the main was depressurized for repair, local regulatory requirements must be met before placing it back into service. At a minimum, if the main was depressurized and there is a potential for contamination, precautions should be taken to ensure that the pipeline is disinfected or otherwise restored to convey safe drinking water. See AWWA Standard C651—Disinfecting Water Mains (ANSI/AWWA C651-14) for guidance. Local drinking water regulations have requirements for issuing and rescinding boil water advisories or boil water orders and for bacteriological sampling and testing. A utility needs to conform to these requirements.

In the event of a main break, designated employees must be prepared to notify and interact with customers, the community, other utilities, public officials, regulators, and the media. When a leak occurs on a customer-controlled service line, the utility must have a system in place to notify the customer of their responsibility to repair the service line and ensure follow-up actions are completed. When leaks and breaks are caused by contractors or other utilities, additional procedures and employees may be involved.

The utility establishes targets to ensure timely investigation and repair of leaks and breaks. Response times can be affected by many factors including dispatch efficiency, labor and equipment availability and expertise, accessibility and condition of valves, traffic conditions, and location. Many utilities dispatch a first responder to investigate the event and secure the area and then dispatch a repair crew only after confirming the leak or break location and severity. Each utility establishes its own processes and procedures that optimize response time within cost and resource limits.

Performance Assessment Process for Main Break Frequency

To complete the performance assessment for main break frequency, the utility collects, organizes, and analyzes performance data for managing reported breaks and leaks on utility controlled transmission and distribution pipes. Although data for break and leak events on service lines are also important, these events are not included in this performance assessment. Although main break data are commonly collected by many utilities, some utilities will not have the type, quantity, or quality of data needed to demonstrate optimization and attempt to correlate main break occurrence with other parameters. A multiyear effort may be needed to both gather the necessary information, and to determine accurately the system status compared with the optimization goals.

The definitions associated with leaks and breaks used in this self-assessment are adopted from the 3rd edition of the AWWA M36 publication, *Water Audits and Loss Control Programs* (AWWA 2016b) and are displayed in Table 2-11.

Table 2-11. Leak and break definitions

Term	Definition
Reported Leaks and Breaks	Leaks or breaks that come to the attention of the water utility as reported by customers, traffic authorities, or any outside party due to their visible and or disruptive nature. A nonsurfacing break initially reported as loss of pressure by a customer is an example of a nonsurfacing reported leak. Events that can be inferred by alerts by supervisory control and data acquisition systems can be labeled as reported breaks. Water utilities respond to reported breaks in a reactive mode, often under emergency conditions.
Unreported Leaks and Breaks	Leaks or breaks that escape public knowledge and are only identified through the active leak detection work performed by the water utility. A leak detection survey is the traditional means of identifying such leaks, but the use of district metered areas (DMA) and leak noise monitors represent emerging technologies that offer improved capabilities to detect unreported leakage. A comprehensive leak detection program is a component of optimized distribution system operation. Water utilities respond to unreported leaks in a planned mode, and can often schedule detection and repair work with minimal disruption to customer service or damage to surrounding infrastructure. Some unreported breaks can have high water loss or cause serious subsurface damage, yet are otherwise unnoticed.
Background Leakage	Tiny weeps and seeps at pipe joints and service connection piping joints that are sonically undetectable. Such tiny leaks can be numerous and widespread in a given distribution system, but are not readily detectable individually. Excessive background leakage can typically be mitigated by pressure management and infrastructure renewal.

Optimization Goal: Main Breaks and Leaks

The Partnership's numeric optimization goal for main breaks is supported by data in several industry publications, including AWWA Benchmarking Survey data and the 1995 Water Research Foundation publication, *Distribution System Performance Evaluation* (Deb et al. 1995). The AWWA publication *Benchmarking Performance Indicators for Water and Wastewater Utilities Survey* (conducted regularly by AWWA) indicates top quartile performance ranging from 14.9–21.7 reported breaks and leaks per 100 miles of utility owned pipeline, annually, based on a survey of about 150 utilities. Consideration of industry data with respect to main break performance, led to the development of the Partnership's optimization goals for main breaks and leaks. The optimization goals for main breaks and leaks are displayed in Table 2-12.

Utilities that have met the performance goal should strive to continue to reduce reported break and leak events. Utilities can show progress toward optimization by completing the main breaks assessment annually and exhibiting a continuing annual decrease in main break frequency. Upon reaching the performance goal, utilities demonstrate, annually, that they are continuing to meet the goal. Utilities may wish to evaluate administrative policies, with respect to main replacement, in conjunction with this review.

Utilities that have significant data limitations should complete the performance assessment using whatever leak and break data is available. In this case, acquiring better and more comprehensive data should be included as part of the utility's performance improvement action plan, so that a more representative main break evaluation can be conducted in the future. Once data deficiencies are addressed, utilities should collect the data for a year and then reassess their performance. It may take several years of main break data collection for the utility to develop a performance baseline and to fully complete the assessment process. It may take even longer to demonstrate optimization.

Some utilities may have most of the data needed for the performance assessment but have difficulty extracting and compiling all of the required information, especially if an electronic database does not exist. These utilities should perform an initial evaluation and reassess performance at a time at which the utility's data are better organized. Utilities may consider the benefits of including information technology staff on the self-assessment team in order to better facilitate the transfer and communication of main break and other data or to develop a comprehensive database for main break information.

Main Break Locations

Location is not a predetermined parameter because data collection is based on events as they are reported or identified. The location of a main break is, however, a key data element that must be collected for each event. This makes it

Table 2-12. Main breaks and leaks optimization goals

Parameter	Goals
Main Breaks and Leaks	No more than 15 reported breaks and leaks per 100 miles of utility-controlled distribution and transmission piping per year. This includes breaks and leaks on appurtenances such as valves, hydrants, blowoffs, fittings, tapping sleeves and saddles, and the connected tapping valves and corporation stops. It does not include service lines and appurtenances beyond the connecting point at the main, regardless of whether utility-owned or controlled or customer owned or controlled. Reported leaks and breaks include both emergency events (that is, needing immediate response and repair) and nonemergency events (that is, addressed during normally scheduled work hours).
Main Breaks and Leaks (alternative goal)	Declining rolling 5-year Main Break frequency trend (most recent 5 years). Systems that do not meet the Main Break Frequency numerical goal can alternatively demonstrate optimized procedures with a 5-year declining Main Break Frequency trend. Each year, a new 5-year trend is calculated to show that progress is continuing toward the optimization goal.
Data Collection	Optimized systems have in place a comprehensive record system that documents the circumstances of the main break or leak. These records are routinely reviewed to determine preventative actions and possible rehabilitation and replacement programs. Main Break records may include but are not limited to: • Pipe segment • Material • Age • Joint type • Internal and external condition • Location (geographic information system) • Unique identifier • Date event occurred • How leak was identified (reported or unreported) • Asset type • Failure mode
Leak Detection	Optimized systems also have an effective and systematic leak detection program. This program uses system flow, pressure, and site studies to locate leaks. The unreported leaks, identified through leak detection are not added to the main break frequency numbers and are ideally repaired before they become reported leaks that would affect this performance measure. Additional information about leak detection may be accessed in the Water Loss Control section of chapter 3.

possible to associate leaks and breaks with a specific pipe segment or pressure zone. Accurately identifying and recording the location of a main break or leak is important to its repair and analysis. The use of GPS data is common and recommended to most accurately record the location of a main break.

Main Break Data Collection

The primary method of capturing the information associated with main break occurrences is each utility's standard data collection forms (electronic or hardcopy) that are completed by water utility field personnel. Considering the variety of data that can and should be collected associated with break and leak investigation and repair, a form dedicated to leaks and breaks is preferable to a multipurpose form, such as a general work order. For utilities that have limited distribution system records, a leak or break event is an opportunity to collect and document information about a pipe segment, such as material, age, joint type, and internal and external condition, which can then be added to the utility's asset management database, if such a database is maintained.

The consistency and completeness of data capture is important for the collection of accurate main break data. All utility employees that record main break data should be trained in proper use of the data collection forms and/or software to help ensure data accuracy. Accurate data are imperative for accurately analyzing the data and drawing valid and relevant conclusions. It is important to clearly define all data fields and use check boxes or dropdown lists as much as possible to ensure consistent data entry. Data collection should be reviewed by supervisory staff at least monthly, but preferably weekly, to ensure field personnel are properly collecting the data and managing leak investigation and response activities.

Because data collection is based on field observation and manual data entry, instrumentation is not an essential element in the accuracy and completeness of leak and break performance data. However, when flow monitoring and SCADA equipment are used to identify hidden leaks, the associated instrumentation should be inspected and maintained routinely, recalibrated appropriately following the manufacturer's recommendations, and regularly verified to ensure maximum effectiveness and performance accuracy. Instrumentation found to be out of calibration at any time should be recalibrated and verified, so that accurate performance can be ensured. Calibration, verification, and maintenance procedures should conform to the manufacturer's recommendation. Acoustic monitoring equipment does not typically require calibration, because monitors typically fail by producing either a continuous signal or no signal. When this occurs, the monitors should be assessed to confirm whether physical equipment failure has occurred. Monitors may also fail are to detect leaks because of poor acoustic transmission (for example from change in pipe materials, presence of repair clamps, or noise).

Assessment Software

Main break data can play an important role in supporting future capital improvement projects and in risk assessment analysis. Therefore, it is important that main break data collected and recorded are as accurate as possible.

The Partnership's main break data collection software is intended to be used by utilities on an annual basis. This frequency provides an opportunity for a thorough analysis of leak and break data. Where sufficient prior data exists, the annual analysis should include trending of data over multiple years, preferably over a minimum of a five-year period (five years is the minimum period required to demonstrate a five-year decreasing trend in main break frequency). Utilities may submit additional years of main break data, to illustrate longer term trends, if it is available. Attention to seasonal variations and periods of extreme weather or other nonroutine events should be considered in evaluating trends. Main break data may also be correlated with other distribution system parameters such as pressure or pipeline age. The need for evaluating data in multiple ways is critical to optimizing many components of distribution system operations. Instructions for software use are provided by the user guides embedded in the software.

The Partnership's main break data collection software performs the calculations and presentation of data that are needed for completing the self-assessment process for main breaks; however, utilities may use a variety of other software and database tools to collect, extract, and compile main break data. Utilities are encouraged to enter field data into a database as soon as possible after the event occurs. If the utility does not currently use an electronic database for its leak and break data, the assessment software can be used to capture data for individual events. However, if the utility already has a more robust database, it should continue to be used and summary data can be extracted annually and copied into the PSW software. Submission of the data generated by the Partnership software is a required component of the self-assessment process. Additional data from utility databases may be submitted if they support specific self-assessment findings or results. Data capture from acoustic leak detection efforts or directly from the SCADA system may require special software. Most data logging vendors have software protocols to aid in the data transfer process. This specialized software is not essential to using the Partnership's assessment tool, but it is necessary to ensure that leaks and breaks identified through use of SCADA and acoustic leak detection devices are accurately entered into the utility's electronic database.

Although the data collection form should include several key data elements (such as pipe size, material, etc.), only the following occurrence data for the reported leaks and breaks are required for the Partnership's performance assessment, including the following fields, which are entered into the Partnership's main break data collection software:

- Date the event occurred
- Number of occurrences
- Identity of the entity that controls the pipe or asset on which the leak or break occurred (utility-controlled vs. customer-controlled). Only leaks and breaks occurring on utility-controlled pipeline are counted in

calculating the occurrence rate against the performance goal. The leaks and breaks that occur on customer-controlled pipe typically relate to customer-owned service lines, which are not included in this evaluation. These events should be tracked by the utility but are not evaluated as a component of the Partnership's self-assessment process.

- Total miles of main, from the utility's inventory of distribution system assets.

- How leak or break was identified ("reported" vs. "unreported"). Only reported leaks and breaks as defined previously (those reported by employee, customer, or public) are counted in calculating the occurrence rate against the performance goal. The unreported leaks, detected with acoustic monitoring equipment or other investigative methods, should be tracked but are not evaluated in this assessment. The Partnership's main break data collection software has the ability to distinguish between reported and unreported breaks.

- Type of asset that is leaking—this includes whether the leak or break occurred on a main, service, valve, tapping saddle, hydrant, or other location. Leaks on most asset types, except service lines, are counted in calculating the occurrence rate against the performance goal.

The software provides the opportunity to enter additional details for each main break occurrence, such as location, failure mode, soil type at the main break location, and information about water loss quantity, although this additional information is not required in order to complete the self-assessment process.

Data Analysis and Outputs

The Partnership's main break data collection software provides a tabular and graphical summary of main break performance, both over time and relative to the Partnership's optimization goal for main breaks and leaks. The graphical summary of main breaks includes a trend line to illustrate the utility's progress toward optimization over the period captured in the software. This information is used in conjunction with additional leak and break information, collected by the utility, to address the self-assessment questions presented in this section of the guide. An example of the graphical output generated by the Partnership's main break data collection software is presented in Figure 2-4.

Status

To determine the optimization status of the distribution system with respect to main breaks and leaks, develop and review the results of the Partnership's main break data collection software results, along with the utility's operational

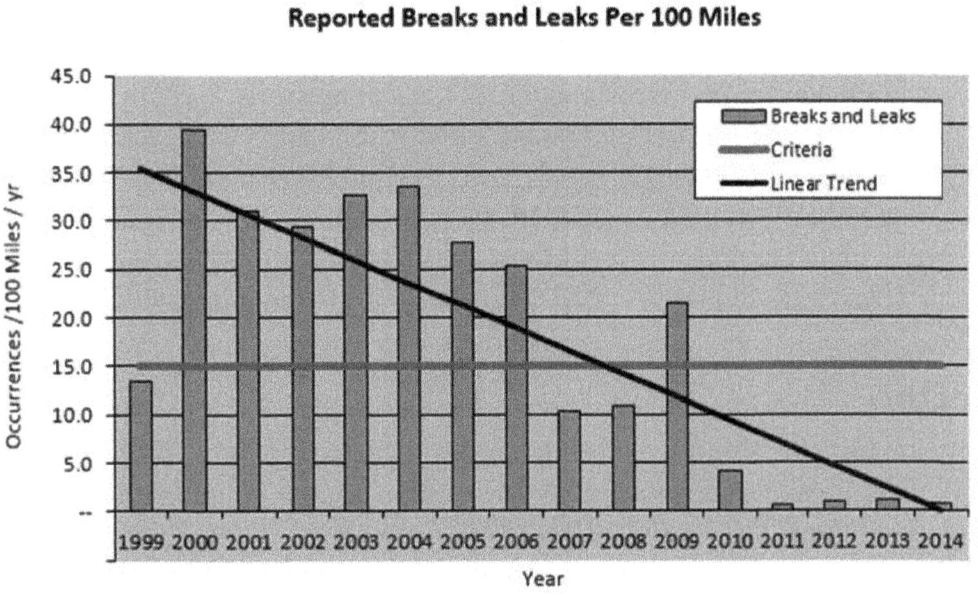

Reported Breaks and Leaks Per 100 Miles

Figure 2-4. Summary graph of main break performance generated by the Partnership's main break data collection software

practices related to main breaks and leaks. Based on this information, review the following items:

- Does the system meet the main break goals?
  - ▲ No more than 15 reported breaks and leaks per 100 miles of utility-controlled distribution and transmission piping per year. This includes breaks and leaks on appurtenances such as valves, hydrants, blowoffs, fittings, tapping sleeves or saddles and the connected tapping valves and corporation stops. It does <u>not</u> include service lines and appurtenances beyond the connecting point at the main, regardless of whether utility-owned or controlled or customer owned or controlled. Reported leaks and breaks include both emergency events (those needing immediate response and repair) and nonemergency events (addressed during normally scheduled work hours).
  - ▲ Systems that do not meet the Main Break Frequency goal can demonstrate optimized procedures with a declining 5-Year Main Break Frequency trend. Each year, a new 5-year trend is calculated to show progress is continuing toward the optimization goal.
  - ▲ Utilities that achieve the performance goal should strive to continue to reduce reported break and leak events. Because it may take many years for some utilities to achieve the performance goal, these utilities can demonstrate progress by completing the assessment annually

and exhibiting a continuing downward trend year-over-year until reaching the performance goal.

- Are main breaks correlated to variations in pressure? Is there an opportunity to reduce system pressures while still meeting all pressure management goals to reduce the frequency of main breaks?

- Has the utility established SOPs to ensure appropriate and timely response to main break/depressurization events? Is all appropriate staff trained in these procedures? Are SOPs regularly reviewed and updated?

Action

If limitations are identified during the status review of the distribution system's main break performance assessment, steps should be initiated to mitigate these limitations. This section offers some alternatives to consider for addressing identified deficiencies. Utility staff should assess the benefits of these recommendations, as well as discuss additional utility-specific actions for continuous improvement.

- If the results of the performance assessment show that the system consistently meets the main break goals, then the performance is excellent. Any activities identified in the remaining chapters of the self-assessment that impact main break frequency will reflect further refinement of operation or administration practices rather than result in significant main break frequency improvements.

- Even if the utility meets the performance goals, an assessment should also be made to ensure the consistent performance is not related solely to a relatively new system that rarely challenges the operators' process control capabilities. New distribution system assets can lead to operator and administrator complacency. Consequently, performance may suffer during rare times when unusual circumstances challenge staff process control capabilities.

- It is important to evaluate the data to identify long-term trends that may affect main break and leak frequency.

- If pressure has been demonstrated to correlate with main breaks, consider addressing pressure management issues to evaluate their impact on main breaks.

  ▲ Examples of addressing pressure issues may include reducing pressures (even a small decrease in pressure has been associated with reducing main break frequency) or reducing the intensity of pressure fluctuations. When making pressure changes, be sure that all regulatory requirements for minimum pressure are met.

▲ If pressure does not correlate with main breaks, consider whether other parameters correlate more directly with main breaks, and address these parameters as possible. Some parameters may be able to be impacted by operational changes, whereas others, such as seasonal variability, are best addressed through utility knowledge and preparation.

- If pressure stabilization procedures are not applied, consider taking the following steps to resolve:

  ▲ Develop or update standard operating procedures with respect to pressure stabilization in response to main breaks.

  ▲ Ensure that field staff are trained in proper main break stabilization procedures.

Performance-Limiting Factor Summary

Table 2-13 summarizes factors related to main break frequency that may be limiting optimized distribution system performance. Indicate whether these factors are Optimized and Documented, Partially Optimized, or Not Optimized. Factors

Table 2-13. Main break assessment

Self-assessment category	Questions for gauging optimization status	Optimization status			Comments
		Optimized and documented	Partially optimized	Not optimized	
Main Breaks	Does the system meet the main break goals?				Submit main break frequency data collection spreadsheet.
	Are main breaks correlated to variations in pressure? Is there an opportunity to reduce system pressures while still meeting all pressure management goals to reduce the frequency of main breaks?				
	Has the utility established SOPs to ensure appropriate and timely response to main break/depressurization events? Is all appropriate staff trained in these procedures? Are SOPs regularly reviewed and updated?				

that are not fully optimized will be prioritized in chapter 7, Continuous Improvement Plan, and an action plan should be developed to address these areas.

References

Table 2-14. References for main break management

Reference	Description
AWWA, 2015. G200-15, *Distribution Systems Operation and Management.* Denver, Colo.: AWWA.	Establishes a standard with minimum requirements for tracking main breaks and maintaining pipeline maintenance records and replacement programs.
AWWA, 2014. ANSI/AWWA C651-14. *Disinfecting Water Mains.* Denver, Colo.: AWWA.	Describes methods of disinfecting newly installed water mains and mains that have been repaired.
Deb, A.K.; Grablutz, F.; Hasit, Y.; Snyder, J.; Loganathan, G.; & Agbenowsi, N., 2002. AwwaRF #459. Prioritizing Water Main Replacement and Rehabilitation. Denver, Colo.: AWWA Research Foundation.	Provides sample main break data collection form and recommendations for performance tracking and analysis.
Deb, A.K.; Hasit, Y.J.; Grablutz, F.M.; Weston Inc., R.F., 1995. AwwaRF #804. Distribution System Performance Evaluation. Denver, Colo.: AWWA Research Foundation.	Provides main break data collection guidelines, an example form, and application to distribution system performance criteria and infrastructure improvement programs.
Kirmeyer, G.J.;Thomure, T.M.; Rahman, R.; Marie, J.L.; LeChevallier, M.W.; Yang, J.; Hughes, D.M.; & Schneider, O., 2014. WRF #4307. Effective Microbial Control Strategies for Main Breaks and Depressurization. Denver, Colo.: Water Research Foundation.	Categorizes main breaks into four types, depending on the severity of the event, and describes appropriate responses to mitigate microbial risks, based on the event categorization.

PERFORMANCE IMPROVEMENT VARIABLES

(content in this chapter is largely from
Water Research Foundation project 4109, Friedman et al. 2010)

Introduction

Discussion in chapter 2 focused largely on defining distribution system integrity with regard to 3 critical components/areas, and representing those areas in terms of key optimization criteria.

Disinfectant residual, pressure management, and main break frequency are explicitly used as the key criteria/indicator for the status of the respective critical integrity components (water quality integrity, hydraulic integrity, physical integrity) in an assessment of a utility distribution system. The three key optimization indicators listed in Table 3-1 are addressed, in depth, throughout chapter 2 of this guide. An expanded discussion of the significance of each indicator is provided in chapter 2, along with self-assessment questions and tables of references. Because of the thorough discussion and supporting information provided for these key areas in chapter 2, the information will not be repeated here. A synopsis of the rationale for the use of these criteria as indicators follows:

Disinfectant Residual—used as indicator for water quality integrity. Numerous references exist (refer to chapter 2) that address the critical nature of maintaining an adequate disinfectant residual in the distribution system in order to protect public health and ensure drinking water quality. The inability of a system to adequately maintain disinfectant levels throughout the distribution system

Table 3-1. Key distribution system optimization indicators

Water Quality	Disinfectant residual (maintain adequate disinfectant residual)
Hydraulic	Pressure (maintain adequate pressure)
Physical	Main breaks (reduce main break frequency)

provides an opportunity for optimization and improved water quality. Several operational variables have the potential to significantly impact disinfectant residual concentrations in the distribution system. Efforts to improve the status of these variables will, in turn, yield benefit in the performance data for this key performance indicator.

Pressure Management—used as the key indicator for the hydraulic integrity of the system. Maintenance of adequate pressure constitutes an important barrier for ensuring both water quality and hydraulic integrity within distribution systems. Loss of pressure, typically below 20 psi, constitutes a public health risk and provides an opportunity for the introduction of contamination into the system. Inadequate pressure management strategies can significantly impact water quality and can also have detrimental effects on mains and other critical system infrastructure. Efforts to improve the status of operational variables affecting pressure management will yield benefits in the overall optimization status of this key performance indicator.

Main Break Frequency—Physical integrity is indicated by an assessment of main break frequency. Efforts to reduce main break frequency will improve the physical integrity of the distribution system, as will efforts to improve the status of operational variables impacting main break frequency. Several variables, such as hydrant and valve maintenance, pipeline rehabilitation and replacement, internal and external corrosion, and storage tank maintenance and operation can significantly affect the frequency of main breaks within the distribution system.

Performance Improvement Variables

Chapter 3 focuses on several operational variables, or Performance Improvement Variables, that significantly affect or influence the performance status of the key optimization criteria. Efforts to focus on the optimization of these operating variables will improve system performance in the key integrity areas listed.

Table 3-2. Performance improvement variables impacting key optimization criteria

Cross-Connection Control and Back-flow Prevention Program	External Corrosion Control	Nitrification	Storage Facility Operations and Maintenance
Customer Complaints	Flushing	Pipeline Installation, Rehabilitation, and Replacement	Water Age, Modeling
Disinfection By-Products	Maintaining Hydrants, Valves, and Blowoffs	Post-precipitation, Inorganic Accumulation	Water Loss Control
Energy Management	Internal Corrosion	Security and Online Monitoring	Water Quality Sampling and Response

The Performance Improvement Variables addressed within this chapter are listed next.

Each of the operating variables listed are important; however, it is not necessary to optimize every one to gain improvements in system performance. Utilities must evaluate the most appropriate variables to target for the situations encountered during self-assessment process. This chapter provides information to assess the optimization status of each Performance Improvement Variable. Variables that are not optimized presently provide opportunities for improvement in system performance and may be addressed through the development of an optimization action plan for the highest priority items, as described in greater detail in chapter 7. Self-assessment questions are included for each Performance Improvement Variable. Utilities are encouraged to assess all applicable Performance Improvement Variables. If a Performance Improvement Variable is not applicable to the utility, a brief explanation of why this is the case should be included in the self-assessment completion report.

It is important to note that the Performance Improvement Variables included in this chapter are not intended to be a comprehensive list of *all* the possible processes that may improve distribution system operation. The factors listed here are the most common and critical variables used to improve distribution system operation. They may be used as a starting point to encourage additional discussion and evaluation during the self-assessment process. Experience and literature research have demonstrated that improvements in these operational areas often result in water quality improvement and increased safety and security. *Utilities completing the self-assessment process should include an evaluation of any other appropriate performance improvement factors, even if they are not explicitly included in this chapter.*

The discussion for each Performance Improvement Variable included in this chapter includes the now-familiar subdivisions:

- *Understanding,*
- *Status,*
- *Action,* and
- *Performance-limiting Factors.*

These are presented in a different way than the previous chapters. The *Understanding* section appears immediately following the Performance Improvement Variable topic heading. This section provides an overview of the topic as it relates to optimized performance of the distribution system. Additional references are listed for each topic, so a deeper understanding can be gained to help during the self-assessment process and when developing and implementing optimization activities. Following the *Understanding* section and table of references, the *Status* section provides a list of self-assessment questions for each topic area. These self-assessment questions are also listed within the *Performance-limiting Factors* Summary Table, included at the end of each topic section. The self-assessment team addresses these questions to assess the optimization status (Optimized and Documented, Partially Optimized, or Not Optimized) and identify performance-limiting factors, so that an action plan to address them may be developed and implemented. The *Status* questions are limited to only a few significant issues that should serve to illuminate important opportunities for improvement. Self-assessment teams may consider accessing the additional references when developing their performance improvement plan (described in chapter 7).

Action

General action(s) required to address performance-limiting factors are common to all variables and, therefore, are discussed here in the introductory section of the chapter. Actions relating to specific Performance Improvement Variables are addressed in the applicable sections of this chapter.

Optimized and Documented

If the response to a self-assessment question indicates that performance is Optimized and Documented, further follow-up actions are not required. The self-assessment completion report should contain appropriate explanation and/or data to support the selected optimization status. In this case, the system consistently meets the optimization goals for one or more of the criteria, and performance, in this area, is excellent.

Any subsequent activities identified that impact the specific self-assessment area will then reflect further refinement of operation or administration practices and may not result in significant performance improvements. Even if the utility meets the performance goals, an assessment should still be completed to ensure that the consistent performance is not related solely to a situation that rarely challenges the operators' process control capabilities, as this can lead to operator and administrator complacency. Consequently, performance can suffer on the occasion when unusual circumstances challenge staff process control capabilities.

Partially Optimized/Not Optimized

If the response to a self-assessment question indicates that performance is Partially Optimized or Not Optimized, an opportunity for improvement may exist. Improving the status of these variables will likely lead to better system performance. It is recommended to develop an action plan to improve the performance in areas for which the self-assessment questions have responses of Partially

Optimized or Not Optimized. Potential steps to consider in the development of an action plan include:

- Identifying and prioritizing the variables that affect each performance optimization criterion.
- Developing an optimization plan for each variable selected for improvement.
    - ▲ Include measurement of the results as a key component to ensure the plan is producing the needed improvement.
- If the system is unable to meet any of the performance goals, for the Performance Improvement Variables selected for improvement consider/assess:
    - ▲ Possible deficiencies with the process control program (see chapter 5).
    - ▲ Possible inadequacies in system design for current operation (see chapter 4).
    - ▲ Evaluating the administrative support level (see chapter 6).
    - ▲ Developing and implementing a performance improvement plan (see chapter 7).

CROSS-CONNECTION CONTROL AND BACKFLOW PREVENTION PROGRAM

Understanding

An effective cross-connection control program involves installation, testing, and continuing maintenance of backflow prevention devices to avoid potential for contamination within drinking water distribution systems. AWWA Manual M14 provides detailed cross-connection control program descriptions (AWWA 2015). Additional references are provided in Table 3-3. Local cross connection control and backflow prevention regulations may apply and should be used when appropriate. The self-assessment questions for this Performance Improvement Variable are included in Table 3-4. Components of a distribution system that is optimized with respect to cross-connection control and backflow prevention are listed below.

Components of an Optimized System for Cross-Connection Control and Backflow Prevention

An optimized cross-connection control and backflow prevention program (Foundation for Cross-Connection Control and Hydraulic Research 2012) includes:

1. The program should have legal authority and administrative responsibility. If there are no state or local regulations, the program regulations

should be as stringent, at a minimum, as those provided in AWWA Manual M14. Coordinate the program with local authorities (such as building, plumbing, and health officials). Provide adequate notice to affected stakeholders and offer regular meetings with them.

2. The cross-connection control program budget, funding, and staff resources must be adequate. Consider administrative costs as well as resources for program implementation.

3. The program includes record-keeping and data management systems. Record-keeping is essential to the success of a cross-connection control program. Specialized computer database programs for cross-connection control programs are available. Key information in the database includes:

   a. The number of customers requiring a device,
   b. The number of devices per customer account,
   c. Type of device and test reports for the device, and
   d. Any notices sent to the customers.

4. Employee qualifications, training, and evaluation procedures are part of a complete program.

5. The program must clearly define the service policy with proper risk assessment (for example, decide if the focus of the program is containment or premises isolation, and whether to focus on all customers or only high-hazard nonresidential customers).

6. Identify customer classifications (such as industrial, commercial, residential, or recycled water) and the type of hazard (such as customers with underground sprinkler systems, private fire protection systems, auxiliary water supplies, and pressure booster).

7. Establish minimum standards for acceptable backflow-prevention assemblies and field test equipment (for example *Double Check Valve Backflow-Prevention Assembly*, ANSI/AWWA C510-07).

8. Ensure proper training of employees, either through qualified internal sources or through credible external providers. Verify certification for cross-connection installers and testers. Backflow Preventer Assembly field testers must be certified.

9. Provide a public education program. Explain the reasons for the program and its benefits. Key information to consider including in educational materials: definitions/descriptions of backflow and cross-connections, the health impact that can occur from backflow events and/or cross-connections, descriptions of situations requiring backflow prevention, requirements for devices and testing schedules, types of

backflow prevention devices, documentation requirements (e.g., permits, certification testing).

10. Install, test, and maintain backflow-prevention assemblies where required. Require testing for all assemblies at least annually.

11. Enforce action for noncompliance (for example, shutoff service for high-risk connections), exercise the shutoff service action with caution, and consult legal assistance if the action is causing substantial loss for a customer.

12. Prepare a backflow-prevention program report, backflow incident response plan, and the program's key measures for effectiveness evaluation.

Table 3-3. References for cross-connection control and backflow-prevention programs

Reference	Description
AWWA, 2007. ANSI/AWWA C510-07, *Double-Check Valve Backflow-Prevention Assembly*. Denver, Colo.: AWWA.	The AWWA standard for backflow-prevention devices.
AWWA, 2015. M14—*Backflow Prevention and Cross-Connection Control: Recommended Practices*, 4th ed. Denver, Colo.: AWWA.	Provides guidance for operating a cross-connection control program. Specifies backflow-prevention devices, industry standards, and criteria for testing backflow-prevention assemblies.
Foundation for Cross-Connection Control and Hydraulic Research, 2012. *Manual of Cross-Connection Control, Tenth Edition*. Los Angeles, Calif.: University of Southern California.	Provides guidance for operating a cross-connection control program. Specifies backflow-prevention devices, industry standards, and criteria for testing backflow prevention assemblies.
Lee, J.J.; Schwartz, P.; Sylvester, P.; Crane, L.; Haw, J.; Chang, H.; & Kwon, H.J., 2003. AwwaRF #90928F. Impacts of Cross-Connections in North American Water Supplies. Denver, Colo.: AWWA Research Foundation.	Conducted a survey of cross-connection and backflow-prevention practices. Provided examples of how abrupt flow changes could trigger low distribution system pressures.
Schneider, O.D.; Bukhari, Z.; Hughes, D.; Fleming, K.; LeChevallier, M.W.; Schwartz, P.; Sylvester, P.; & Lee, J.J., 2009. WRF #3022. Cross-Connection and Backflow Vulnerability: Monitoring and Detection. Denver, Colo.: Water Research Foundation.	Monitors backflow events using reverse flow sensing water meters and demonstrates that low level backflow occurs at 1-2% of the locations per month.
USEPA, 2002b. Potential Contamination Due to Cross-Connections and Backflow and Associated Health Risks USEPA white paper. Washington, D.C.: US Environmental Protection Agency.	Reviews state requirements for cross-connection control, case studies of backflow events, and risks. Emphasizes that training and certification are important.

Status

The status of the cross-connection control and backflow prevention program as it relates to achieving a desired level of distribution system performance may be assessed by reviewing the following self-assessment questions. The self-assessment team is not limited to these questions and may consider discussing additional topics related to cross-connection control and backflow prevention that may assist in identifying and addressing performance-limiting factors.

- Does the utility have authority to enforce backflow regulations (local, state/provincial, federal or other)?

- Does the utility have a comprehensive list of locations where backflow prevention devices are required and testing and verification schedules and results?

- Does the utility provide training and verify certification for cross-connection installers and testers?

- Does the utility have a comprehensive backflow prevention program that includes all the elements in AWWA Manual M14 and meets all local, state/provincial, or other applicable requirements?

Action

If areas of the cross-connection control and backflow prevention program are considered to have a status of Partially Optimized or Not Optimized, and this performance improvement variable has been prioritized and selected for improvement, the following steps are recommended to be considered in the development of an action plan to improve performance in this area.

- If the utility does not have the authority to enforce backflow regulations, consider developing a working relationship with the enforcement entity, such that the potential impact of cross-connection control and backflow prevention is fully understood. Consider building utility awareness of enforcement procedures to ensure they are consistent with the water utility's objectives.

- If the utility does not have a summary list of the number and location of backflow prevention devices located in the system, consider taking action to build and develop such a list over time. This list may include testing and verification schedules as well as location and physical details.

- If appropriate training is not provided by the utility, consider verifying through the training providers that all cross-connection installers and testers are appropriately trained and certified.

- If the utility has not established a comprehensive backflow prevention program, consider developing a program, as outlined in AWWA Manual M14, and ensure that the program meets all applicable regulatory requirements.

Performance-Limiting Factors Summary

Table 3-4 summarizes factors related to cross-connection control and backflow prevention that may be limiting optimized performance of the distribution system. Check whether these factors are Optimized and Documented, Partially Optimized, or Not Optimized. Factors identified as Partially Optimized or Not Optimized will be prioritized in chapter 7, Identification and Prioritization of Performance-Limiting Factors, where an action plan can be developed and implemented that allows for optimization of parameters selected for improvement.

Table 3-4. Cross-connection control and backflow prevention assessment

Self-assessment category	Questions for gauging optimization status	Optimization Status			
		Optimized and documented	Partially optimized	Not optimized	Comments
Enforcement Authority	Does the utility have authority to enforce backflow regulations (local, state/provincial, federal, or other)?				
Backflow Protection Device Locations	Does the utility have a comprehensive list of locations where backflow prevention devices are required and testing and verification schedules and results?				
Backflow Protection Device Installation and Testing	Does the utility provide training and verify certification for cross-connection installers and testers?				
Backflow Program	Does the utility have a comprehensive backflow prevention program that includes all the elements in AWWA Manual M14 and meets all local, state/provincial, or other applicable requirements?				

CUSTOMER COMPLAINTS

Understanding

Customer complaints are important measures that can be used to assess distribution system optimization. These contacts with customers are important opportunities for utilities to gain key information that is useful to ensure quality of service, safety of water, security information, and to optimize overall performance. Customer complaint response is important for distribution system optimization, because this interaction often results in a conclusion about the customer's satisfaction with service. In some areas, reporting procedures and content may be specified under state or local regulatory requirements. Additional references related to customer complaints can be found in Table 3-5. The self-assessment questions for this Performance Improvement Variable are included within this section and summarized in Table 3-6. Components of a distribution system that is optimized with respect to cross-connection control and backflow prevention are listed below.

Components of an Optimized System for Customer Complaint Response

1. Set up a complaint response program meeting the requirements of state regulators, public utility commissions, health departments, and other regulatory officials.

2. Budget for continuing administrative, staffing, and technology costs. Systems with large service areas and diverse operations typically require cross-departmental communications for customer complaint handling.

3. Document and track customer calls by type, date, location, caller, response, results of response follow-up, and details of remedial actions taken. Complaint histories should be reviewed by area with routine operations reviews and capital investment planning programs.

4. Quickly and effectively receive and prioritize customer complaint calls to identify and respond to urgent calls in a timely manner. A 24-hour contact system facilitates customer access to qualified response personnel. Responders must be trained to recognize situations where there is potential for serious health risk or property damage. Cross-connections should be investigated as potential causes of water quality irregularities.

5. Calls about adverse taste, odor, or color should be investigated and resolved to determine the possibility of more serious issues.

6. Provide appropriate customer information to supplement complaint investigations. Such information may be provided by customer call numbers, websites, fliers, mailings, postings, door hangers, print- or electronic-media.

7. Keep up-to-date contact information for critical- or special-needs customers and agencies (such as bulk customers, bulk suppliers, health service providers, schools, emergency response entities, regulators, and news media).

8. Optimized distribution systems should have few complaints. AWWA Benchmarking Surveys consistently show the best performing systems have less than 2.5 Technical Quality Complaints per 1,000 customer accounts, on an annual basis. This represents a target goal for which optimized systems may strive.

Technical quality complaints are defined as complaints that are directly related to the core services of the utility. This includes complaints associated with water quality, taste, odor, appearance, pressure, main breaks, and disruptions of water service.

Status

The status of customer complaints as it relates to achieving a desired level of distribution system performance may be assessed by reviewing the following self-assessment questions. The self-assessment team is not limited to these questions and may consider discussing additional topics related to customer complaints that may assist in identifying and addressing performance-limiting factors.

- Are technical customer complaints recorded and tracked separately from billing and general information inquiries?
- Is customer water quality complaint response time tracked?
- Does the number of annual technical water quality complaints compare favorably with the AWWA benchmarking survey results (2.5 per 1000 customer accounts per year)? If not, is there a plan to reduce this number?
  - ▲ Calculate and report the number of technical water quality complaints per 1000 customer accounts for inclusion in the self-assessment completion report. Consider describing any utility-specific ways that customer complaints are categorized.

Action

If areas of the utility's handling of customer complaints are considered to have a status of Partially Optimized or Not Optimized, and this performance improvement variable has been prioritized and selected for improvement, the following

Table 3-5   References for effective customer complaint and inquiry response

Reference	Description
AWWA, 2015. G200-15, Distribution Systems Operation and Management. Denver, Colo.: AWWA.	Specifies differentiation of complaints; action plan to address complaints with trained personnel.
AWWA, 2005. *Water Distribution System Assessment Workbook*. Smith, C., editor. Denver, Colo.: AWWA.	Recommends recording discolored water calls and determining causes; identify and remediate low flow area/correct at plant; flush known deposits; conduct follow-up testing. Check forms are included.
AWWA, 2016e. Benchmarking Performance Indicators for Water and Wastewater Utilities: 2016 Edition. Denver, Colo.: AWWA. ISBN 9781625761972.	AWWA's Benchmarking Surveys provide an annual summary of key utility performance indicators for drinking water, wastewater, and combined utilities.
Besner, M.C.; Gauthier, V.; Trepanier, M.; Martel, K.; Prévost, M., 2007. Assessing the Effect of Distribution System O&M on Water Quality. *Journal AWWA*, 99(11):77-91.	Article includes references of heterotrophic plate count bacteria, and customer complaints.
Burlingame, G.A., 2007. Ratcheting Up Lab Response to Water Quality Warnings. *Opflow*, 33(2):24-27.	Article describes data handling, and management of water quality and customer complaints to identify capabilities/limitations, expand safety programs, and sampling, and managing water quality complaints.
Deb, A.K.; Hasit, Y.J.; Grablutz, F.M.; Weston Inc., R.F., 1995. AwwaRF #804. Distribution System Performance Evaluation. Denver, Colo.: AWWA Research Foundation.	Report recommends three DS performance criteria based on customers' needs: Adequacy (water quantity, quality); Dependability (DS capability for consistency); Efficiency (water, energy resources).
Kirmeyer, G.J.; Friedman, M.; Clement, J.; Sandvig, A.; Noran, P.; Martel, K.; Smith, D.; LeChevallier, M.; Volk, C.; Antoun, E.; Hiltebrand, D.; Dyksen, J.; Cushing, R., 2000. AwwaRF #90798. Guidance Manual for Maintaining Distribution System Water Quality. Denver, Colo.: AWWA Research Foundation.	Describes potential causes of WQ degradation with recommended O&M and design to prevent or remediate. Field validation studies conducted in 19 systems.
NASTB, 2006. Drinking Water Distribution Systems: Assessing and Reducing Risks. Washington, D.C.: National Research Council Report.	Recommends that areas with repeat customer complaints about color, taste, or odor; (e.g., dead end mains and storage facilities) be flushed.
Whelton, A.J.; Dietrich, A.M.; Gallagher, D.L.; Roberson, J.A., 2007. Using Customer Feedback for Improved Water Quality and Infrastructure Monitoring. *Journal AWWA*, 99(11):62-76.	Research found that effective capture and analysis of customer feedback about water quality, quantity, aesthetics, illness, and security can be an effective early warning system for monitoring water quality.

steps are recommended to be considered in the development of an action plan to improve performance in this area.

- If customer complaints are not tracked, consider developing a formal mechanism through which customer complaints may be accurately tracked and categorized.
  - ▲ Consider tracking the date and time the complaint was both received and resolved, to determine response time. This data may be used to assess complaint response procedures.
  - ▲ Similarly, if technical water quality complaints are not differentiated from more general complaints, consider developing a mechanism to better categorize the complaints received.
  - ▲ Consider mapping customer complaints to identify potential geographic patterns in occurrence and to more readily correlate with related parameters.
- If the number of annual technical water quality complaints significantly exceeds AWWA's benchmarking survey results, consider developing a plan to further evaluate customer complaints that may allow for the root causes of the complaints to be identified.
  - ▲ Based on the distribution system events most frequently resulting in customer complaints, a plan may be developed to address the most common sources of customer complaints. Ensure that customer education plays an appropriate role in this process.
- If response time to customer water quality complaints is not tracked, assess the benefits of implementing a mechanism that allows for tracking and improvement of overall response time. Consider developing goals for this parameter and ensure that utility staff is aware of, and trained in, all appropriate response procedures and requirements.

Performance-Limiting Factor Summary

Table 3-6 summarizes factors related to customer complaints that may be limiting optimized performance of the distribution system. Check whether these factors are Optimized and Documented, Partially Optimized, or Not Optimized. Factors identified as Partially Optimized or Not Optimized will be prioritized in chapter 7, Identification and Prioritization of Performance-Limiting Factors, where an action plan can be developed and implemented that allows for optimization of parameters selected for improvement.

Table 3-6.   Customer complaints

Self-assessment category	Questions for gauging optimization status	Optimized and documented	Partially optimized	Not optimized	Comments
Customer Complaint Monitoring and Response	Are technical customer complaints recorded and tracked separately from billing and general information inquiries?				
	Is customer water quality complaint response time tracked?				
Technical Quality Complaint Goal	Does the number of annual technical water quality complaints compare favorably with the AWWA benchmarking survey results (2.5 per 1000 customer accounts per year)? If not, is there a plan to reduce this number?				Calculate and report the number of technicalcomplaints/1000 customer accounts

DISINFECTION BY-PRODUCTS

Understanding

The regulated total trihalomethanes (TTHMs) andfive haloacetic acids (HAA5) represent the chlorinated disinfection by-products (DBPs) of primary concern (USEPA 2006d). The process of optimizing DBP concentrations through the water treatment plant and distribution systems is complex. Many factors influence DBP formation and composition, such as:

- The disinfectant
- Disinfectant dosage and residual concentrations maintained
- Point of disinfectant application
- The concentration and characteristics of natural organic matter precursors at point of disinfection application
- pH
- Temperature
- Bromide concentration
- Disinfectant contact time
- Water retention time in system piping and storage, and
- Booster chlorination, if used (USEPA 2006c-d, 2007)

Components of an Optimized System for DBP Control

DBPs can undergo growth, decay, and transformation in the distribution system (AWWA/EES 2002a; Pereira 2004; USEPA 2006c, Routt et al. 2009). Ideally, a fully optimized system will manage and balance all of the factors listed above to control DBP formation. This ensures meeting the DBP goals both at both the point of entry (POE) and throughout the system. Often, maximum HAA5 and TTHM levels do not occur simultaneously and do not occur at the same sites. Formation of HAA5 is typically more rapid than formation of TTHM and occurs to a greater extent closer to the treatment plant than in the distribution system. Higher pH favors TTHM formation, and lower pH favors HAA5 formation. Utilities should identify options to optimize treatment and distribution system practices to reduce DBP potential formation.

Once the water has entered the distribution system, DBP control may be optimized through implementation of the appropriate following practices:

- Reducing water age
- Blending source waters or purchased finished water with low DBP precursor concentrations (the presence of other parameters, such as bromide, in the blended water source should be evaluated if this option is selected)
- Storage tank aeration to reduce the levels of more volatile TTHM species such as chloroform (this is not necessarily effective for all DBP species)
- Reducing the concentration of free chlorine applied at entry points, and adding chlorine boosters in the system
- Switching to chloramine (AWWA 2006a; USEPA 2006c-d)
- Flushing

Chlorine residual and TTHM levels should vary inversely (as TTHM levels increase, residuals decrease). HAA5s increase through the system, except where degradation is occurring. HAA5s can undergo both chemical and biological degradation. An example of chemical degradation is the breakdown of brominated trihaloacetic acids into THMs (Zhang 2002). An example of biodegradation is mono- and di-haloacetic acids metabolized by heterotrophic bacteria (Bayless 2008). A result of this is that HAA5 concentrations can be highest near the system entry points where heterotrophic plate count (HPC) levels are lowest. HAA5s can also be proportionately higher in colder water when microbial metabolic rates are lower.

Areas of the system with elevated DBP concentrations, elevated levels of heterotrophic bacteria, and low disinfectant residuals may be in need of improved system turnover or booster disinfection (USEPA 2006d). Utilities can trend DBP and disinfectant residual levels across the system to identify patterns and irregularities (Routt et al. 2009). The levels of TTHM and HAA5 formed are much lower with chloramine compared to free chlorine (Pierra 2004, USEPA 2006c). Many systems with high precursor levels in their finished waters have switched to chloramination for secondary disinfection.

A good first step in any system DBP evaluation is to determine the DBP levels entering the system. If these levels are already high, DBP formation may need to be addressed through treatment modifications. For utilities that receive water as a consecutive system or from a wholesaler, it is important to work closely with the supplier regarding DBP levels in the delivered water. Next, comparison of entry point DBP levels with levels occurring in the distribution system can reveal formation patterns. As discussed previously, there are many variables besides water age that can affect DBP formation. Studies may reveal opportunities to implement measures to optimize DBP levels, along with the performance of related parameters. Simultaneous compliance can be complex, with many variables to consider; therefore, it is always advisable to approach treatment and/ or distribution system changes with caution. Piloting on some scale, if at all possible, is recommended before implementing full-scale changes to assess their potential impact on all areas of water quality.

Temperature can be an important factor in chlorine demand reactions, DBP formation, and biological growth. Some systems will experience peak DBP formation levels during the warmest water temperature months. Other systems may experience peak formation as a function of seasonal variations in precursor levels. Reducing water age can also result in less DBP formation (AWWA/EES 2002a; Grayman 2004).

Additional references about DBPs can be found in Table 3-7. The self-assessment questions for this Performance Improvement Variable are included in Table 3-8.

Status

The status of DBP control as it relates to achieving a desired level of distribution system performance may be assessed by reviewing the following self-assessment questions. The self-assessment team is not limited to these questions and may consider discussing additional topics related to customer complaints that may assist in identifying and addressing performance-limiting factors.

- Do TTHM and HAA5 distribution system test results satisfy the regulatory requirements?
    - ▲ The USEPA maximum contaminant levels (MCLs) are ≤ 80 µg/L for TTHM and ≤ 60 µg/L for HAA5, calculated as a locational running annual average. Regulatory requirements may vary in other countries, and utilities are encouraged to evaluate the status of this question with respect to their local regulatory requirements.

Table 3-7. References for disinfection by-product control

Reference	Description
AWWA and Economic and Engineering Services. 2002a. *Effects of Water Age on Distribution System Water Quality.*	USEPA Total Coliform Rule issue paper. Water age impacts DBPs and DBPS and can be indicators of water age.
AWWA. 2006a. M20, *Water Chlorination and Chloramination Practices and Principles, Second Edition.* Denver, CO: AWWA.	Manual describes chemical properties, disinfection mechanisms, feed rates, and strategies/techniques to minimize disinfection byproducts.
AWWA. 2012. *Standard Methods for Examination of Water and Wastewater.* 22nd Edition. E.W. Rice, R.B. Baird, A.D. Eaton, L.S. Clesceri, editors. ISBN 978-0-87553-013-0	Extensive listing of USEPA-approved analytical methods including DBPs, disinfectant residuals, and microbials.
Bayless, W. and R.C. Andrews. 2008. Biodegradation of six haloacetic acids in drinking water. *Journal of Water Health* 6(1):15-22.	Research documented loss of mono and di HAAs exposed to biofilms.
Grayman, W.M., L.A. Rossman, C. Arnold, R.A. Deininger, C. Smith, J.F. Smith, and R. Schnipke. 2000. AwwaRF #260. Water Quality Modeling of Distribution System Storage Facilities.	Spatial temperature difference in storage and the system can suggest stratification in storage tanks and degradation of water quality in the distribution system.
Pereira, V., H.S. Weinberg, and P.C. Singer. 2004. Temporal and spatial variability of DBPs in a chloraminated distribution system. *Journal AWWA* 96(11):91-102.	Article explains factors that affect DBPs in chloraminated systems.
Routt, J. C., M. Sekhar, M. Friedman. 2009. Optimizing Drinking Water Distribution Systems for Disinfection and Disinfection Byproducts. *Conference Proceedings Disinfection 2009.* Atlanta, Ga: WEF/AWWA.	Article explains USEPA and Water Research Foundation research in progress involving a variety of utilities working to optimize systems for DDBPs.
USEPA. 2001. Controlling Disinfection By-Products and Microbial Contaminants in Drinking Water. USEPA/600/R-01/110.	This 2001 document summarized (microbial/DBP control) technology research by USEPA since the publication of USEPA's prior 1981 treatment technology document.
USEPA. 2006. Initial Distribution System Evaluation Guidance Manual for the Final Stage 2 Disinfectants and Disinfection Byproducts Rule. EPA 815-B-06-002. Washington, D.C.	USEPA manual describes sampling and system evaluation approaches to assess DBP levels throughout water distribution systems.
USEPA. 2006. Stage 2 Disinfectants and Disinfection Byproducts: Final Rule. *Fed. Reg.* 71:2:388.	USEPA federal regulations for disinfectants and disinfectant byproducts.
USEPA. 2007. Simultaneous Compliance Guidance Manual for the Long Term 2 and Stage 2 DBP Rules. Office of Water (4601. EPA 815-R-07-017).	EPA manual provides integrated references and utility case studies to illustrate potential effects of system or operations changes to meet microbial/DBP control requirements.
Zhang, X., R.A. Minear. 2002. Decomposition of trihaloacetic acids and formation of the corresponding trihalomethanes in drinking water. *Water Res.* 36(14):3665-3673.	The two mixed chlorobromo species, BDCAA and DBCAA, were found to decompose to form BDCM and DBCM, respectively, via a decarboxylation pathway.

- Are system DBP concentration trends monitored, and is there a plan to maintain or reduce the current levels while maintaining adequate disinfectant residual?

Action

If areas of the utility's DBP control practices are considered to have a status of Partially Optimized or Not Optimized, and this performance improvement variable has been prioritized and selected for improvement, the following steps are recommended to be considered in the development of an action plan to improve performance in this area.

- If TTHM or HAA5 distribution system concentrations do not meet regulatory requirements, consider evaluating treatment and/or distribution system practices that may serve to reduce DBP formation.

  ▲ In this process, consider the factors that can impact DBP formation, such as chlorine residual levels, concentration, and structure of natural organic matter in the water, pH, temperature, water age, and other distribution system operational factors, and other influencing water quality characteristics, such as bromide levels.

  ▲ If the key drivers for DBP formation can be identified, this may help to identify the next steps to take to minimize DBP formation in the distribution system, as described previously in this section.

  ▲ Utility staff should carefully assess the benefits of piloting potential DBP reduction options before full-scale implementation, so that their overall impact on water quality may be fully assessed.

- If DBP concentrations are not currently regularly monitored and trended, establish procedures to sample for DBPs in high-risk areas of the system (this may involve separate sampling sites for TTHM and HAA5), collect samples, and record results in a such a way that they can be monitored, trended, and evaluated on a regular basis.

Performance-Limiting Factor Summary

Table 3-8 summarizes factors related to DBP control that may be limiting optimized performance of the distribution system. Check whether these factors are Optimized and Documented, Partially Optimized, or Not Optimized. Factors identified as Partially Optimized or Not Optimized will be prioritized in chapter 7, Identification and Prioritization of Performance-Limiting Factors, where an action plan can be developed and implemented that allows for optimization of parameters selected for improvement.

Table 3-8. Disinfection by-product control

Self-assessment category	Questions for gauging optimization status	Optimization Status			
		Optimized and documented	Partially optimized	Not optimized	Comments
DBP goals	Do TTHM and HAA5 distribution system test results satisfy the regulatory requirements? (USEPA MCLs are 80 µg/L for TTHM and 60 µg/L for HAA5, calculated as a locational running annual average—other countries may vary)				
DBP monitoring and trending	Are system DBP concentration trends monitored, and is there a plan to maintain or reduce the current levels while maintaining adequate disinfectant residual?				Assess the results shown on the disinfection performance spreadsheet

ENERGY MANAGEMENT

Understanding

According to the Electric Power Research Institute (EPRI 2002), approximately 4 percent of the electricity produced in the United States is for water and wastewater treatment and distribution. In California, water-related energy use consumes 19 percent of the state's electricity (Klein 2005). Current and pending regulations to improve water quality or use poor source waters are projected to increase energy usage (Chang et al. 2008). Although electric use varies by the source of supply (surface vs. groundwater) and terrain (some areas can be served by gravity), the typical system expends approximately 90 percent of its energy use in distribution system pumping; therefore, focusing on distribution system energy management can be an effective way to reduce energy consumption and associated energy costs. It is important that energy management initiatives be fully integrated with the utility's water quality and operational goals.

Components of an Optimized System for Energy Management

Several research studies performed by the Water Research Foundation, the New York State Energy Research Development Authority, Southern California Edison, and the European Commission share the following similar findings:

- Historically, energy audits have found old pumps to have an average wire-to-water efficiency averaging 55 percent (Conlon et al. 1999).

- The average pump efficiency decreases 10-25 percent over its lifetime (Easton Consultants 1995).

- European Commission pump efficiency reports indicate that large pumps lose 5 percent of their efficiency in the first 5 years of operation (European Commission 2001).

- New York State Energy Research Development Authority reports that the average 10-year-old pump can typically gain 10 percent efficiency by sandblasting all wetted parts and resealing with a ceramic epoxy. Another 10 percent can be saved by trimming the impeller and replacing worn parts, throat bushings, wear rings, and seals (Maier 2008).

- Hydraulic Institute standards for new, large pump installations (> 500 HP with horizontal split case pumps) target a 78–82 percent wire-to-water efficiency (European Commission 2001).

- Hydraulic Institute standards for new, small pump installations (< 500 HP with horizontal split case pumps) target a 65–70 percent wire-to-water efficiency (European Commission 2001).

The best efficiency point for the pump is the point of highest efficiency on the pump curve. Conducting an annual wire-to-water measurement can determine the pump efficiency (Rishel 2002). The best option is to collect individual pump flow and power use, and if this is done in real-time (through the supervisory control and data acquisition system), the system can constantly monitor and adjust the pumping efficiency. To do this, the system will need the real-time power use from the individual pump, the pump curve for each pump, pump flow and pressure data, and storage tank level data. An excellent wire-to-water efficiency at peak demand would be > 80 percent and a good value would be 74–80 percent. Typical actions to improve pump efficiency could include:

- Replacing inefficient pumps and motors
- Replacing pump impellers and bowls
- Trimming pump impellers
- Refurbishing existing pumps with new low-friction coatings
- Eliminating pump discharge throttling and by-passing
- Adding variable speed drives, where appropriate
- Replacing inefficient adjustable speed drives

Eliminating distribution system flow restrictions from small mains or corroded pipes can increase pump efficiency by decreasing the pump head. Reducing or optimizing system pressures can also increase energy efficiency. Other options for reducing energy costs include changing significant pump operation activities to times when the energy cost is lower. For example, in areas where there is "time-of-day" energy pricing, shifting pumping to off-peak times and using system storage during peak hours can reduce energy costs. These actions may serve to reduce energy costs, although they will not necessarily reduce energy consumption. Improvements in water loss control can also reduce energy consumption and costs. A reduction in system leakage results in less water produced for the same customer population, thereby reducing the energy associated with water production. Implementing improved pressure management can reduce energy consumption by both lowering the demand for pumping, as well as potentially reducing distribution system water losses.

Additional references regarding Energy Management can be found in Table 3-9. The self-assessment questions for this Performance Improvement Variable are included in Table 3-10.

Status

The status of energy management as it relates to achieving a desired level of distribution system performance may be assessed by reviewing the following self-assessment questions. The self-assessment team is not limited to these questions and may consider discussing additional topics related to energy management that may assist in identifying and addressing performance-limiting factors and improving performance and efficiency.

- If pressure is managed to reduce energy usage, does the utility monitor the effect on stability, insuring that appropriate pressure ranges are maintained?

- Are all major distribution system pumps routinely tested for efficiency? Are there targets for maintenance or replacement based on efficiency?

- Has a hydraulic surge analysis been performed and addressed where required?

Action

If areas of the utility's distribution system energy management practices are considered to have a status of Partially Optimized or Not Optimized, and this performance improvement variable has been prioritized and selected for improvement, the following steps are recommended to be considered in the development of an action plan to improve performance in this area.

- If pressure management is used as a means of reducing energy usage, and the stability of distribution system pressures is not able to be verified, consider the increased use of pressure monitoring to evaluate pressure

Table 3-9. References for energy management

Reference	Description
Burton, F., 1996. *Water and Wastewater Industries Characteristics and Energy Management Opportunities.* Report CR-106941. Palo Alto, Calif.: Electric Power Research Institute.	Provides information on energy use by plant size and process. Also includes information on energy intensity of new processes.
Carlson, S. & Walburger. A., 2007. AwwaRF #91201. Energy Index Development for Benchmarking Water and Wastewater Utilities. Denver, Colo.: AWWA Research Foundation.	Report provides a tool by which utilities can benchmark their energy use using a multi-parameter method similar to the USEPA's ENERGY STAR rating system for buildings.
Chang, Y.; Reardon, D.J.; Kwan, P.; Boyd, G.; Brant, J.; Rakness, K.L.; & Furukawa, D., 2008. WRF #3056. Evaluation of Dynamic Energy Consumption of Advanced Water and Wastewater Treatment Systems. Denver, Colo.: Water Research Foundation.	Evaluates factors that affect energy consumption by advanced water/wastewater treatment technologies and identifies energy optimization opportunities.
Conlon, T.; Weisbrod, G.; & Samiullah, S., 1999. We've Been Testing Water Pumps for Years—Has Their Efficiency Changed? *Proceedings of the 1999 ACEEE Summer Study on Energy Efficiency in Industry.*	Reports the results of over 28,000 pump tests performed by Southern California Edison between 1990 and 1997. Provides data on existing pump efficiency and trends in pump efficiency by pump type, size, and applications.
Dufresne, L., 2016. *Energy Management for Water Utilities.* Denver, Colo.: WWA.	Provides an overview of key steps in developing and implementing an energy management program, including benchmarking, auditing, and economic analysis.
Easton Consultants, 1995. Strategies to Promote Energy-Efficiency Motors Systems in North America's OEM Markets. Stamford, Ct.	Widely cited report that pumps lose 10-25% efficiency over their lifetimes.
EPRI, 2002. *Water and Sustainability, Volume 4.* US Electricity Consumption for Water Supply and Treatment—The Next Half Century. Report: 1006787. Palo Alto, Calif.: Electric Power Research Institute.	Details energy use by US water utilities. Forecasts demand.
European Commission, 2001. Study on Improving the Energy Efficiency of Pumps. *AEAT-6559/v5.*1.	Includes curves for theoretically attainable, maximum practically attainable, catalog mean, Hydraulic Institute "Large Pumps," and Hydraulic Institute "ANSI/API."
Klein, G., 2005. *California's Water Energy Relationship.* Final Staff Report, CEC-700-2005-011-SF. Sacramento, Calif.: California Energy Commission.	Details the water/energy relationship in California.
Maier, P.; White, R.; Connell, S.; Kroll, K.; King, C.; Hanley, G.; & Metzger, R., 2008. *Energy Savings through Pump Refurbishment and Coating.* Pumps & Systems.	Reports interim results of 5-year study that is evaluating the impact of refurbishment and coatings on pump efficiency. Each of the 18 pumps will be tested regularly over the 5-year duration of the project.

Reference	Description
Rishel, J.B., 2002. *Water Pumps and Pumping Systems.* New York, N.Y.: McGraw-Hill Professional.	Textbook on conducting pump wire-to-water efficiency measurements.
USDOE Energy Efficiency and Renewable Energy. Best Practices. Pumping Tip Sheets.	Provides Tip Sheets on how to test pump efficiency and all aspects of improving pumping efficiency, including adjustable speed drive applications, pump selection, pipe sizing, and impeller trimming/replacing.

at key locations throughout the system. Refer to chapter 2 for additional information regarding pressure monitoring and management.

- If distribution system pumps are not routinely tested for efficiency, consider the benefits of developing a program for pump efficiency testing.

  - Such a program should set appropriate targets for pump efficiency.

  - Consider incorporating pump efficiency test results as a driver for key preventative maintenance and/or pump replacement activities.

- If a hydraulic surge analysis has not been performed, consider completing such an assessment and implementing appropriate actions based on the outcome.

Performance-Limiting Factor Summary

Table 3-10 summarizes factors related to energy management that may be limiting optimized performance of the distribution system. Check whether these factors are Optimized and Documented, Partially Optimized, or Not Optimized. Factors identified as Partially Optimized or Not Optimized will be prioritized in chapter 7, Identification and Prioritization of Performance-Limiting Factors, where an action plan can be developed and implemented that allows for optimization of parameters selected for improvement.

Table 3-10. Energy management

Self-assessment category	Questions for gauging optimization status	Optimization status			Comments
		Optimized and documented	Partially optimized	Not optimized	
Pressure Management	If pressure is managed to reduce energy usage does the utility monitor the effect on stability insuring that appropriate pressure ranges are maintained?				
Pump Efficiency	Are all major distribution system pumps routinely tested for efficiency? Are there targets for maintenance or replacement based on efficiency?				
Pump Operation	Has a hydraulic surge analysis been performed and addressed where required?				

EXTERNAL CORROSION CONTROL

Understanding

Corrosion results in premature deterioration and loss of pipe material, leading to leakage and pipe failure. The most common causes of external corrosion on iron water mains include:

- Corrosive soil environment
- Galvanic corrosion by coupling to dissimilar metals
- Corrosion from stray currents caused by electric lines or an impressed current cathodic protection system (typically used on gas mains).

Additional references regarding external corrosion control can be found in Table 3-11. The self-assessment questions for this Performance Improvement Variable are included in Table 3-12. Internal corrosion control is another Performance Improvement Variable that is covered later in this chapter.

Components of an Optimized System for External Corrosion Control

Tracking of Leaks and Breaks

Information collected from the investigation of main breaks and leaks can be useful in determining the root cause of the event. This information can help to influence future actions that the utility may take with respect to pipe rehabilitation and/or replacement, such as pipe material selection, the use of cathodic protection, or the implementation of additional external corrosion prevention measures.

It can be important to collect this pertinent information at the time a main break or leak occurs and store the data in a location and format in which it may be easily referenced in the future. In addition to the field data described in the main breaks management section of this guide, when corrosion is suspected, it is important to also document the following items:

- The type of leak or break (joint leak, longitudinal split, circumferential break)
- Pipe depth
- Form of corrosion (corrosive soil, stray current, galvanic)
- Depth and diameter of pitting
- Pipe thickness remaining
- Position of the failure on the pipe (top, bottom, side)

A visual assessment of surrounding soil type (sand, clay) and knowledge of soil saturation conditions before the break or proximity to water table should be recorded to assist with assessing the cause of the failure. The number of failures from reported and unreported breaks and leaks resulting from external corrosion should be tracked to determine if they are increasing or decreasing over time. The system should routinely review trends in leaks and breaks (consider evaluating failures by pipe class and/or location) to identify corrosion hot spots and gauge the effectiveness of current practices.

Investigation and Analysis

It is recommended that the utility have a knowledgeable employee or consultant who is responsible for managing and investigating leaks and breaks where corrosion is suspected. The investigation should include observation of soil characteristics representative of the bedding conditions near the break, as well as any appropriate soils testing. For evaluating existing pipes, the process may include conducting soils surveys and stray current testing or excavating and inspecting existing pipes. The soils may be analyzed for parameters including pH, redox potential, sulfide, resistivity, and electrical conductivity.

Locating the water main should include the consideration of road surfaces, gas mains, impressed current rectifiers or other sources of electric current, and the presence of dissimilar metals near the failure site. If a pipe section is removed to make the repair, structural testing may be helpful, especially if there is no clear cause of failure.

A mechanistic model, although complex and difficult to calibrate, may be most suitable for prioritizing renewal of cast iron water mains (Deb 2002). Makar (2005) also provides some recent testing results to correlate failure to pitting size, loading, and soil conditions. AWWA Manual of Practice M27—External Corrosion for Infrastructure Sustainability (AWWA 2013a) is another good source of information for the evaluation of external corrosion.

Mitigation of Corrosion Effects

The utility should have standards in place for the protection of new ductile iron or steel water mains. Corrosion mitigation practices for existing pipe may include a retrofit program for cathodic protection of the main or replacement with materials better suited for the field conditions. Research by the National Research Council of Canada has indicated that sacrificial anodes at corrosion-related main breaks do reduce the frequency of subsequent breaks, but an organized program for corrosion mitigation can be even more effective.

It may be helpful to conduct soils surveys in conjunction with the selection and design of new mains, particularly for transmission mains that will have serious consequences from failure or those that expand into new and unfamiliar areas. Design standards and materials selection criteria should consider corrosion-resistant piping materials, plastic wraps and coatings, insulators, bonded joints, installation and monitoring of cathodic protection systems, or other methods to control external corrosion. AWWA Manual of Practice M27 (AWWA 2013a) is a good source of information. Additional information may be obtained from pipe manufacturers' associations, such as the Ductile Iron Pipe Research Association.

Status

The status of external corrosion control practices as related to achieving a desired level of distribution system performance may be assessed by reviewing the following self-assessment questions. The self-assessment team is not limited to these questions and may consider discussing additional topics related to external corrosion control that may assist in identifying and addressing performance-limiting factors. Note that internal corrosion control is a separate Performance Improvement Variable covered later in this chapter.

- Are pipes inspected and sampled whenever they are exposed?
  - ▲ Times when pipes are exposed provide an opportunity to collect physical samples of pipe (if appropriate), take any appropriate surficial measurements, or document pipe condition using photographs.
- Are corrosion observations recorded during all main break and leak repairs?

- Are soils always tested before installing metal pipes or materials?
- Is corrosion protection installed when materials and environmental conditions warrant such measures?

Table 3-11. References for the development of external corrosion control programs

Reference	Description
AWWA, 2015. G200-15, Distribution Systems Operation and Management. Denver, Colo.: AWWA.	Establishes a standard with minimum requirements for monitoring external corrosion.
AWWA, 2005. *Water Distribution System Assessment Workbook*. Smith, C., editor. Denver, Colo.: AWWA.	Provides checklists and additional references for evaluating and controlling external corrosion of buried pipelines in water distribution systems.
AWWA, 2013. *M27—External Corrosion for Infrastructure Sustainability*, 3rd ed. Denver, Colo.: AWWA.	Provides background information, guidelines, and techniques to understand the chemistry and mechanics of corrosion, evaluate corrosion potential, and control corrosion of pipelines.
Deb, A.K.; Grablutz, F.; Hasit, Y.; Snyder, J.; Loganathan, G.; & Agbenowsi, N., 2002. AwwaRF #459. Prioritizing Water Main Replacement and Rehabilitation. Denver, Colo.: AWWA Research Foundation.	Provides methods for analyzing main break data and prioritizing pipeline renewal.
Makar, J.; Rogge, R.; McDonald, S.; & Tesfamariam, S., 2005. AwwaRF #91053. The Effect of Corrosion Pitting on Circumferential Failures in Grey Cast Iron Pipes. Denver, Colo.: AWWA Research Foundation.	Provides results and recommendations from testing of pit cast and spun cast grey iron pipe under various loading and corrosion conditions.

Action

If areas of the utility's distribution system external corrosion control practices are considered to have a status of Partially Optimized or Not Optimized, and this performance improvement variable has been prioritized and selected for improvement, the following steps are recommended to be considered in the development of an action plan to improve performance in this area.

- If pipes are not inspected and sampled whenever they are exposed, consider creating standard operating procedures (SOPs) that allow distribution system staff to take advantage of the opportunity to gather as much information as possible about pipe condition during the limited times when pipes are exposed, including during main break and leak repairs.
  - ▲ Associated with this, consider the benefits of developing a tool or form that allows field staff to readily document corrosion conditions, ensuring that this information is accessible for future evaluation.

- If soils are not routinely tested before installing metal pipes or materials, consider developing procedures that incorporate this step. Even if soil conditions are generally uniform, the potential exists for localized conditions that may affect soil corrosivity.

- If corrosion protection is not installed when materials and environmental conditions warrant such measures, consider the benefits of implementing corrosion prevention measures, such as cathodic protection or the installation of plastic wraps or coatings.

Performance-Limiting Factor Summary

Table 3-12 summarizes factors related to external corrosion control that may be limiting optimized performance of the distribution system. Check whether these factors are Optimized and Documented, Partially Optimized, or Not Optimized. Factors identified as Partially Optimized or Not Optimized will be prioritized in chapter 7, Identification and Prioritization of Performance-Limiting Factors, where an action plan can be developed and implemented that allows for optimization of parameters selected for improvement.

Table 3-12. External corrosion

Self-assessment category	Questions for gauging optimization status	Optimization status			
		Optimized and documented	Partially optimized	Not optimized	Comments
Pipe Inspection	Are pipes inspected and sampled whenever they are exposed?				
Main Break and Leak Records	Are corrosion observations recorded during all main break and leak repairs?				
Soil Studies	Are soils always tested before installing metal pipes or materials?				
Corrosion Protection	Is corrosion protection installed when materials and environmental conditions warrant such measures?				

FLUSHING

Understanding

System flushing is performed to clean distribution pipelines of sediment, impurities, and biofilm. Flushing can also be used to achieve other water quality objectives such as increasing turnover to reduce water age, increase residual chlorine concentrations, decrease DBP levels, and potentially resolve discolored water complaints or taste and odor problems. Flushing programs may be systematic for the entire distribution system or localized to resolve problems in specific pipe segments. A unidirectional flushing program may be the optimal solution to address a variety of distribution system water quality issues (Hasit et al. 1999). In a unidirectional flushing program, specific valves are closed and hydrants opened so the water flows from clean areas with sufficient velocity to clean the system (Antoun et al. 1999). Although flushing can be a beneficial distribution system practice in many cases, it may be minimized in more arid areas where water conservation is of critical importance.

Partnership for Safe Water subscribers are required to submit a corrective flushing SOP as a component of the self-assessment completion report. The development of a flushing SOP is encouraged, even for distribution system that may not flush on a routine basis, so that distribution system staff is prepared in the event of an emergency situation that may require flushing.

Additional references related to distribution system flushing can be found in Table 3-13. The self-assessment questions for this Performance Improvement Variable are included in Table 3-14.

Components of an Optimized System for Distribution System Flushing

Systematic and Localized Flushing Plan

The system should have a written, detailed plan describing the flushing program. Friedman (2005) provides a template for development of such a plan. Many commercially available software packages can aid in developing plans for a unidirectional flushing program. The use of a current distribution system hydraulic model can be helpful in developing a comprehensive flushing plan.

The utility should also develop metrics for assessing water quality in the distribution system to help determine when flushing should be conducted. These criteria could include the number or frequency of water quality complaints received, disinfectant residual concentrations, dissolved oxygen levels, nitrification-related parameters, or high water-age occurrence in low-flow areas (Chadderton et al. 1992). The plan should identify physical assets and location/operation information for hydrants, valves, and blowoffs that may need to be accessed to carry out the flushing process. An asset management or maintenance management program can assist with maintaining an accurate physical inventory, as well as information about operational status of these assets. Procedures should include those associated with flushing and the testing of new or repaired mains.

Metrics to Monitor Flushing

Several metrics should be established to monitor the effectiveness of flushing, including flushing velocity and water quality parameters The flushing velocity should be appropriate for the specific water quality objective. Lower velocities (1 ft/sec or less) may be used when the objective is to move water through the system (to improve disinfectant residual or reduce water age). Higher velocities (2.5-5 ft/sec) may be required to flush sediment or scour biofilms from the system (Friedman et al. 2003). Water pressures should be monitored throughout the area to avoid creating areas of low or negative pressure. Monitoring flows can ensure that fire flow requirements can be achieved. Assessment of water quality parameters (such as disinfectant residual, turbidity, and color) can help to evaluate the effectiveness of flushing, in real time, while flushing is being conducted. Utilities may establish performance goals in these areas to optimize a flushing program.

Public Education

A communication plan should be established to notify critical stakeholders of a planned or continuing flushing program. Internal communications should inform employees and departments that are in contact with the public before the start of a flushing program. Customer service and communications employees should be prepared to inform customers of timing and duration of the program, what to expect during the flushing, and any actions that may be needed to clear services lines upon completion. Customer communication methods may include bill inserts, on-site signage, community cable networks, websites, direct customer letters, external print media, press releases, and social media.

Routine Maintenance

The success of the flushing process is dependent on the physical conditions of the valves, hydrants, and other appurtenances required to carry it out. As a secondary benefit, the utility should have a process in place for assessing the condition of distribution system assets used during the flushing process and a work order system for tracking and prioritizing maintenance that may be needed for these assets. The use of valves, hydrants, and other appurtenances during the flushing process can help distribution system staff to identify any broken or improperly functioning assets.

Dechlorination

The utility flushing plan should consider procedures, such as dechlorination, to neutralize disinfectant residuals to protect waterways and mitigate environmental impacts. In instances where it may be required, dechlorination may be achieved using a variety of practices. Chloraminated systems typically stop ammonia addition and convert to free chlorine to minimize the impacts on nitrogen discharge to receiving streams. Utilities should be aware of and adhere to any local regulatory requirements associated with dechlorination.

Automated Flushing

Automatic flushing systems are an option to address limited areas of deteriorating water quality that result from low water usage. Plans for automated flushing systems include consideration of the location, frequency of operation, and any potential environmental impacts. These systems should be routinely inspected and monitored for operability, effectiveness, and potential impacts on distribution system pressure.

Table 3-13. References for the development of distribution system flushing programs

Reference	Description
Antoun, E.N.; Dyksen, J.E.; & Hiltebrand, D.J., 1999. Unidirectional Flushing: A Powerful Tool. *Journal AWWA*, 91(7):62-71.	An overview of unidirectional flushing, summarizing the key components and potential benefits to control distribution system water quality.
Chadderton, R.A.; Christensen, G.L.; & Henry-Unrath, P., 1992. AwwaRF #515. Implementation and Optimization of Distribution Flushing Programs. Denver, Colo.: AWWA Research Foundation.	A study of complaint records indicated that a spreadsheet could be used to order particular areas within a distribution system by flushing priority. A protocol to initiate and improve flushing programs was developed, based on the concept of various levels of effort for flushing programs.
Friedman, M.; Martel, K.; Hill, A.; Holt, D.; Smith, S.; Ta, T.; Sherwin, C.; Hiltebrand, D.; Pommerenk, P.; Hinedi, Z.; & Camper, A., 2003. AwwaRF #2606. Establishing Site-Specific Flushing Velocities. Denver, Colo.: AWWA Research Foundation.	Utilities that had never flushed before using a unidirectional approach benefited significantly from flushing at a high velocity, ≥ 5 feet per second (fps). Utilities that had recently flushed (within 4-6 years) using a unidirectional approach received little benefit from flushing at a high velocity (≥ 5 fps), versus the lower velocity (2-4 fps). Dead-end flushing is not effective unless the hydrant or blowoff is located at the pipe terminus.
Friedman, M.; Kirmeyer, G.; Pierson, G.; Harrison, S.; Martel, K.; Sandvig, A.; & Hanson, A., 2005. AwwaRF #2875. Development of Distribution System Water Quality Optimization Plans. Denver, Colo.: AWWA Research Foundation.	Developed a template for flushing distribution system of different-sized systems. Program outlined best management practices and audits of flushing programs.
Hasit, Y.J.; DeNadai, A.J.; Gorrill, H.M.; McCammon, S.B.; Raucher, R.S.; & Witcomb, J., 1999. AwwaRF #2605. Cost and Benefit Analysis of Flushing. Denver, Colo.: AWWA Research Foundation.	The software tool developed in this study can be used to guide utilities in developing their optimal flushing strategies.

Status

The status of flushing practices as related to achieving a desired level of distribution system performance may be assessed by reviewing the following self-assessment questions. The self-assessment team is not limited to these questions and may consider discussing additional topics related to flushing that may assist in identifying and addressing performance-limiting factors.

- Has the utility instituted a routine flushing program and documented rationale for flushing practices (or lack of) currently in use?
- Are the following methods incorporated into the flushing program?
  - ▲ Is chlorine residual monitored as part of the flushing program?
  - ▲ Is chlorine residual used as an indicator of when it is acceptable to terminate a flush or used to adjust automatic flusher settings as appropriate?
  - ▲ Are routine flows or flush velocities at least 2.5 ft/sec when removing loose particles?
  - ▲ Are routine flushing velocities at least 5 ft/sec for removal of cohesive and loosely adhered particles and biofilm?
  - ▲ Is dechlorination of flushing water provided where appropriate?
- Is pressure monitored during flushing (manual and automatic) to verify that pressure goals are continuously achieved?

Action

If areas of the utility's distribution system flushing practices are considered to have a status of Partially Optimized or Not Optimized, and this performance improvement variable has been prioritized and selected for improvement, the following steps are recommended to be considered in the development of an action plan to improve performance in this area.

- If the utility has not instituted a routine flushing program, consider the benefits of developing a flushing program.
  - ▲ A comprehensive flushing program should include an overall strategy for flushing, as well as more specific procedures for how flushing is conducted.
  - ▲ Utilities that do not routinely conduct flushing (such as those located in arid climates) are encouraged to develop flushing procedures and track asset conditions so that staff has the knowledge to properly conduct flushing when it may be required.
- If flushing methods have not been established or do not contain velocity, water quality, or dechlorination targets, consider developing SOPs for flushing that incorporate these components.

- If pressure is not routinely monitored during the flushing process, consider monitoring pressures at key locations to verify that flushing is not creating areas of low pressure in the system.

Performance-Limiting Factor Summary

Table 3-14 summarizes factors related to distribution system flushing that may be limiting optimized performance of the distribution system. Check whether these factors are Optimized and Documented, Partially Optimized, or Not Optimized. Factors identified as Partially Optimized or Not Optimized will be prioritized in chapter 7, Identification and Prioritization of Performance-Limiting Factors, where an action plan can be developed and implemented that allows for optimization of parameters selected for improvement.

Table 3-14. Flushing

Self-assessment category	Questions for gauging optimization status	Optimized and documented	Partially optimized	Not optimized	Comments
Routine Flushing Program	Has the utility instituted a routine flushing program and documented rationale for flushing practices (or lack of) currently in use?				
Flushing Methods	Are the following methods incorporated into the flushing program? • Is chlorine residual monitored as part of the flushing program? • Is chlorine residual used as an indicator of when it is acceptable to terminate a flush or used to adjust automatic flusher settings as appropriate? • Are routine flows or flush velocities at least 2.5 ft/sec when removing loose particles? • Are routine flushing velocities at least 5 ft/sec for removal of cohesive and loosely adhered particles and biofilm? • Is dechlorination of flushing water provided where appropriate?				
Pressure Monitoring	Is pressure monitored during flushing (manual and automatic) to verify that pressure goals are continuously achieved?				

MAINTAINING HYDRANTS, VALVES, AND BLOWOFFS

Understanding

An effective maintenance program establishes an asset inventory, condition assessment plan, work management process (tracking, prioritization, and coordination with other work), planning process (analysis, budgeting, and strategic alignment), and communication plan. These assets are typically not operated or needed to deliver water to customers under normal system conditions and, therefore, are often forgotten or neglected. However, these assets become a critical component of distribution system operation and optimization during emergencies or when making system modifications. In an emergency situation, where timing can be critical, it is essential that distribution system staff know where critical assets are located and that these assets be in functional condition. The presence of nonlocatable, inaccessible, or nonfunctional valves can result in extra time, effort, and movement needed to address a main break situation.

Additional references relating to maintaining hydrants, valves, and blowoffs can be found in Table 3-15. The self-assessment questions for this Performance Improvement Variable are included in Table 3-16.

Components of an Optimized System for Maintaining Hydrants, Valves, and Blowoffs

Asset Inventory and Location Information

The utility should have a complete inventory of its hydrants, valves, and blowoffs, which includes associated information about their physical and location attributes. It is best to have this information stored electronically in a database that can be easily sorted, queried, and updated. AWWA M17—Fire Hydrants: Installation, Testing, and Maintenance (AWWA 2016a) and AWWA M44—Distribution Valves: Selection, Installation, Field Testing, and Maintenance (AWWA 2016c) provide guidance, templates, and lists of recommended asset attributes to collect, maintain and update. Many commercially available software packages can aid in the collection, storage, retrieval, sorting, and analysis of asset physical and location data.

The effective management of physical asset and location information for hydrants, valves, and blowoffs can facilitate a utility's ability to locate and operate valves, hydrants, and blowoffs during emergency situations such as main breaks or fires. The ability of distribution system staff to locate and operate these assets also can successfully help to minimize the extent of service disruption to customers, during these events, and minimizes the risk of contamination entering the distribution system during both planned and emergency work.

Condition Assessment

The utility should have a process in place for assessing the condition of distribution system assets. Typically, this is accomplished through regular inspection or operation/exercise of the hydrant, valve, or blowoff. Deficiencies or problems should be documented, and unless the problems can be addressed immediately, the asset should be scheduled for corrective maintenance.

Water flow rate and volume, as well as water quality parameters such as chlorine residual, turbidity, and color, should be measured and recorded during testing or inspection of hydrants and blowoffs. System water pressures should be monitored to avoid producing low or negative pressures. Hydrant flows should be monitored to ensure that fire flow requirements can be achieved when the assets are in use. Monitoring of water quality parameters (such as disinfectant residual, turbidity, color, time required for hydrant flow to run clear) can help to evaluate trends in system performance and assess whether testing and inspection activities are effective. Hydrants found to be nonfunctional should be marked and repaired according to prescribed plans, as well as any applicable regulatory requirements.

Work Management Process

The utility should have a system in place to contain a written or systematic detailed plan describing its maintenance program. AWWA M17 and AWWA M44 provide guidance, templates, and procedures for development of such a plan. The utility should have a work order system in place for tracking and prioritizing maintenance on the assets. When inspecting, exercising, or operating valves, hydrants, and blowoffs, it is critical to document the work performed and the condition of the asset "as found" and "as left" so that broken or improperly functioning assets are identified, repaired, or replaced. Also, the data should be analyzed for trends and patterns that may help to optimize the maintenance program.

Planning

The utility should establish metrics for operability of hydrants, valves, and blowoffs and should develop replacement programs to address groups of assets that have exceeded their useful life or fail to meet required performance levels. Criteria for replacement or rehabilitation could include the number or frequency of failures, the impact of the particular asset's failure on proper system functioning (such as water quality, pressure, flow, and availability), and the cost and benefits of replacement or repair. The planning should be well coordinated with budgeting processes and integrated with other investment programs such as main replacement and extensions of service to new customers to increase the benefit.

Public Education

A communication plan should be established to notify critical stakeholders when planned or corrective maintenance of hydrants, valves, or blowoffs affects the operability of the system or the availability or quality of water in their area. Internal communications should inform employees and departments typically in contact with customers before the start of testing or maintenance work. Customer service and communications employees should be prepared to inform customers of the program's timing and duration, as well as what to expect during the testing or maintenance. Communication methods may include bill inserts, onsite signage, door hangers, community cable networks, Websites, direct customer letters, external print media, press releases, and social media.

Status

The status of hydrant, valve, and blowoff maintenance program practices, as related to achieving a desired level of distribution system performance, may be assessed by reviewing the following self-assessment questions. The self-assessment team is not limited to these questions and may consider discussing additional

Table 3-15. References for development of hydrant, valve, and blowoff maintenance programs

Reference	Description
AWWA, 2015. G200-15, Distribution Systems Operation and Management. Denver, Colo.: AWWA.	Establishes a standard with minimum requirements for programs to test, operate, and maintain valves and hydrants.
AWWA, 2005. *Water Distribution System Assessment Workbook*. Smith, C., editor. Denver, Colo.: AWWA.	Provides checklists for assessing application and maintenance of valves, hydrants, and blowoffs in water distribution systems.
AWWA, 2016. *M17—Fire Hydrants: Installation, Field Testing, and Maintenance*, 5th ed. Denver, Colo.: AWWA.	Provides background information, procedures, forms, and diagrams for the proper installation, inspection, testing, and repair of fire hydrants and guidelines for setting up a maintenance program.
AWWA, 2016. *M44—Distribution Valves: Selection, Installation, Field Testing, and Maintenance*, 3rd ed. Denver, Colo.: AWWA.	Provides background information, procedures, forms, and diagrams for the proper selection, installation, inspection, testing, and repair of valves and guidelines for setting up a maintenance program.
Friedman, M.; Kirmeyer, G.; Pierson, G.; Harrison, S.; Martel, K.; Sandvig, A.;& Hanson, A., 2005. AwwaRF #2875. Development of Distribution System Water Quality Optimization Plans. Denver, Colo.: AWWA Research Foundation.	Provides guidelines and audit procedures. Program outlined best management practices.

topics related to flushing that may assist in identifying and addressing performance-limiting factors.

- Does the system have accurate and current records that document the location and attributes for all valves, hydrants, and blowoffs?

- Are all valves, hydrants, and blowoffs inspected and evaluated on a schedule?

- Are all distribution system main valves and hydrants exercised and tested at least every three years (or more frequently if required by regulation)?

- Are all hydrant repairs scheduled within 24 hours of discovery? Are inoperable hydrants identified immediately and is this communicated to the fire protection authority?

- Does the system control access to hydrants and provide training for proper third-party use?

Action

If areas of the utility's distribution system hydrant, valve, and blowoff maintenance practices are considered to have a status of Partially Optimized or Not Optimized, and this performance improvement variable has been prioritized and selected for improvement, the following steps are recommended to be considered in the development of an action plan to improve performance in this area.

- If the system does not have accurate and current records documenting the location and attributes for all valves, hydrants, and blowoffs, consider developing a system for record-keeping, such as an asset management system.

  - ▲ Refer to chapter 4 for more detailed information about asset management systems and procedures.

  - ▲ Records are recommended to contain information about location, age, operability, and any additional parameters of significance and should be accessible by staff who may need to use the data for evaluation, review, and additional correlation.

  - ▲ Records should be regularly updated to help ensure their accuracy.

- If valves, hydrants, and blowoffs are not regularly evaluated, consider developing a schedule and procedures that allow for periodic evaluation of these assets.

- If valves and hydrants are not regularly exercised, develop a schedule and procedures to establish a regular valve and hydrant exercise program.

    ▲ Consider prioritizing assets to address those most critical to distribution system operation.

    ▲ It is recommended to exercise valves once every three years, or according to local regulatory requirements.

- If procedures do not exist to address inoperable hydrants, it is recommended to develop procedures for reporting inoperable hydrants to the local fire protection authority within 24 hours of their discovery, as well as for scheduling any required repairs within that timeframe.

- If the utility does not currently control access to hydrants, consider the potential benefits of limiting access and providing training to help ensure appropriate third-party use, which may help to minimize risk to the distribution system.

Performance-Limiting Factor Summary

Table 3-16 summarizes factors related to valve, hydrant, and blowoff maintenance that may be limiting optimized performance of the distribution system. Check whether these factors are Optimized and Documented, Partially Optimized, or Not Optimized. Factors identified as Partially Optimized or Not Optimized will be prioritized in chapter 7, Identification and Prioritization of Performance-Limiting Factors, where an action plan can be developed and implemented that allows for optimization of parameters selected for improvement.

Table 3-16. Maintaining valves, hydrants, and blowoffs

Self-assessment category	Questions for gauging optimization	Optimized and documented	Partially optimized	Not optimized	Comments
		Optimization status			
Location Records	Does the system have accurate and current records that document the location and attributes for all valves, hydrants, and blowoffs?				
Inspection and Assessment	Are all valves, hydrants, and blowoffs inspected and evaluated on a schedule?				
Exercise Program	Are all distribution system main valves and hydrants exercised and tested at least every three years (or more frequently if required by regulation)?				Assess the valve exercising program to ensure that the testing frequency goal is met.
Hydrant Repairs	Are all hydrant repairs scheduled within 24 hours of discovery? Are inoperable hydrants identified immediately and is this communicated to the fire protection authority?				
Hydrant Access	Does the system control access to hydrants and provide training for proper third-party use?				

INTERNAL CORROSION

Understanding

Internal corrosion can occur when the chemical, biological, and hydraulic characteristics of drinking water cause internal metallic components to decay. Internal corrosion can lead to two major problems: failure of distribution system pipes, causing leakage/loss of hydraulic capacity and unwanted water quality changes, such as red water resulting from iron corrosion byproducts, and elevated concentration of lead and copper at the tap (AWWA 2011). The result of corrosion can be elevated metal concentrations in the bulk water or metal oxide deposits, pitting, tuberculation, or weakening of piping. A variety of strategies exist for evaluating water's potential for corrosion, as well as for corrosion control. Corrosion control generally refers to the adjustment of water quality at the treatment plant, the application of corrosion inhibitors, or other factors implemented to minimize corrosivity. Corrosion control methods can be system-specific and must be thoroughly tested and evaluated prior to use.

Additional references pertaining to internal corrosion can be found in Table 3-18. The self-assessment questions for this Performance Improvement Variable are included in Table 3-19.

Components of an Optimized System for Internal Corrosion Control

Corrosion is a complex process that occurs on a microscopic level. Corrosion is affected by many water quality parameters such as pH, alkalinity, calcium concentration, dissolved oxygen level, disinfectant concentration, temperature, the presence and concentration of corrosion inhibitors, and other factors. Corrosion indices have been established to evaluate the corrosivity of water. Some of these corrosion indices include indices related to calcium carbonate such as the Langelier Saturation Index (LSI), calcium carbonate precipitation potential (CCPP), Ryznar Saturation Index (RSI), and Aggressiveness Index (AI). Indices related to other water quality parameters include the Larson Index (LI) and chloride to sulfate mass ratio (CSMR). Higher CSMR has been demonstrated to increase lead release into water (Edwards & Triantafyllidou 2007, Nguyen et al. 2011), so when there are treatment or source water changes that may increase CSMR, utilities need to consider its potential impact on lead release. Calculation and application of the different corrosion indices are summarized in Table 3-17.

Another important corrosion indicator is the redox potential, which is a measure of the tendency of a chemical species to acquire electrons and thereby be reduced. Oxidation reduction potential (ORP) is a direct measurement of the

Table 3-17. Corrosion indices calculation and reference range

Corrosion index	Water quality parameters needed	Noncorrosive/scaling reference range
LSI (most commonly used)	pH, $[Ca^{2+}]$, alkalinity, temperature, total dissolved solids (TDS)	–0.5 to 0.5
CCPP	pH, $[Ca^{2+}]$, alkalinity, temperature, TDS	–5 to 0
RSI	pH, $[Ca^{2+}]$, alkalinity, temperature, TDS	≤ 6
AI	pH, $[Ca^{2+}]$, alkalinity	> 12
LI	$[SO_4^{2-}]$, $[Cl^-]$, alkalinity	< 0.5
CSMR	$[SO_4^{2-}]$, $[Cl^-]$	< 0.2

Note: [ ] indicates the concentration of these components; AWWA provides a tool to quickly estimate corrosivity using corrosion indices LSI and RSI, as well as precise estimates using a model. These resources can be accessed at http://www.awwa.org/resources-tools/water-knowledge.aspx.

redox potential. Lytle and Schock (2008) have shown that relatively minor variations in redox potential can directly impact lead/copper mineralogy and stability, which can impact metal solubility and release. Changes in source water and water treatment can greatly impact redox potential, thus impacting corrosivity and lead/copper solubility. For example, it has been demonstrated that changing disinfectant from free chlorine to chloramine can cause increase of lead levels primarily due to the change of ORP (Lytle & Schock 2005, Boyd et al. 2008). Therefore, utilities need to consider the impact on ORP when considering changes in source water and treatment.

Internal corrosion-related monitoring is needed to assess corrosion potential and occurrence in distribution systems and to evaluate the need for or the effectiveness of corrosion control programs. Corrosion monitoring and control are also important to assess and minimize corrosion-related by-product accumulation (such as oxides, hydroxides, hydroxycarbonates, carbonates, or hydroxysulfates [Lytle 2007]), and corrosion control treatment residuals (silicates and phosphates). Allowing corrosion by-products to accumulate can cause aesthetic problems and loss of disinfectant residual from chemical reactivity. Distribution piping that is tuberculated can serve as prime areas for growth of biofilms, which can exert a disinfectant demand. Internal corrosion monitoring and control are also needed to protect system components from structural damage such as pinhole leaks and pipe breaks (USEPA 2007, AWWA 2011).

At the time of this writing, USEPA is currently considering long-term revisions to the Lead and Copper Rule (LCR), which may include substantive changes. Readers are encouraged to access the USEPA Website to obtain current information regarding the regulatory status of the rule. As of late 2016, USEPA's LCR (USEPA 1988b) mandates monitoring of corrosion including the following:

- Lead and copper monitoring at the tap for all systems
- Water quality parameter monitoring for large systems (systems serving > 50,000 persons) and small/medium sized systems that exceed the lead or copper action levels to ensure optimal concentrations are maintained

Corrosion indices, water quality monitoring, dissolved or particulate corrosion products (including lead, copper, iron, etc.) monitoring, combined with pipe observation and coupon and pipe loop studies can help utilities optimize the monitoring of internal corrosion. Corrosivity indicator measurements by physical (coupon weight loss) or electrochemical means (linear polarization probes) can also be useful for determining the potential for corrosion in a distribution system. There are generalized ASTM standards for measuring corrosion rates by coupon weight loss. The results from these test methods are descriptive (mild, moderate, severe), but are not necessarily positively correlated to metal concentrations.

Control of corrosion is made difficult by the wide variety of materials used in distribution systems (such as iron, copper, lead, or cement). An optimized corrosion control strategy for one material may not necessarily be optimal for another. The LCR requires all systems to have optimal corrosion control for lead and copper. It is known that chemical, biological, or hydraulic instability of the bulk water disrupts protective films that may have deposited, resulting in increased corrosivity (USEPA 2003, 2007; Lytle 2007, 2008; AWWA 2011). When a system has long-term changes in treatment processes or is considering switching to a new source, the system typically is required to notify the local regulatory agency, which will then typically review and approve the changes. A system should assess the corrosion potential of the new water and perform corrosion control studies before implementing any changes. The recommended steps to follow are: (1) assess the corrosion potential of the new water by calculating its corrosion indices and comparing it with the existing source, (2) perform bench-scale and pilot-scale studies to demonstrate the impact of the change on corrosion to establish optimal corrosion control strategies, (3) obtain approval from the local regulatory agency before switching to the new water (if required), and (4) conduct distribution system and tap water monitoring after the switch to verify water quality.

Local and federal requirements mandate at-the-tap monitoring for lead and copper and require large systems to establish optimal water quality parameters for minimizing lead and copper solubility. Characteristics of optimized internal corrosion control programs include:

- Maintaining adequate staffing and resources needed to comply with the LCR (USEPA 1988b).

- Establishing written guidelines (standard operating procedures) for control of corrosion-related byproducts (such as iron, color, zinc, taste and odor) in the distribution system to help ensure that the water meets USEPA secondary MCLs for corrosion-related parameters (particularly iron).

- Conducting routine monitoring of corrosion-related water quality parameters (such as pH, alkalinity, conductivity, calcium, and redox potential; and zinc, phosphates, or silicates, if applied for corrosion inhibition). The optimized pH range will be system-specific and dependent on other water quality characteristics, including temperature, alkalinity, hardness, and other ions in the water.

- Evaluating water treatment or source water changes that have the potential to impact water quality (such as pH, alkalinity, disinfection, coagulation, organics, and use of a corrosion inhibitor). This includes studying factors that may lead to revised corrosion control strategies. Regulatory requirements typically apply if corrosion treatment is changed. The local regulatory agency must be notified if changes are planned, and additional lead and copper testing may be required.

- Regularly inspecting the internal condition of piping for tuberculation, pitting, holes, or scaling. Conditions may be correlated with leak detection, C-values, type and composition of scale, and scale integrity.

- Using corrosion monitoring techniques to evaluate corrosion rates for different metals (such as steel, lead, copper, or brass). Tracking and responding to corrosion-related customer complaints. Data should be characterized by complaint type (discolored water, taste and odor, pressure) and mapped to enable detecting areas of greatest concern,

- Utilities receiving water from a wholesaler should assess the corrosivity of the water and design and install equipment to apply its established optimal corrosion control treatment, if necessary.

Table 3-18. References for effective internal corrosion monitoring and control

Reference	Description
ASTM, 2013. G96—90 Standard Guide for Online Monitoring of Corrosion in Plant Equipment (Electrical and Electrochemical Methods). West Conshohocken, Pa.	Method describes procedure for conducting online corrosion monitoring of metals under operating conditions by the use of electrical or electrochemical probe methods.
ASTM, 2015. D2688—15 Standard Test Methods for Corrosivity of Water in the Absence of Heat Transfer (Weight Loss Methods). West Conshohocken, Pa.	Method covers the determination of the corrosivity of water by evaluating pitting and by measuring the weight loss of metal specimens exposed to flowing water from the system.
AWWA, 2011. *M58—Internal Corrosion Control in Water Distribution Systems*. Denver, Colo.: AWWA.	Manual describes best practices and explains concepts related to corrosion-related assessments and control.
Boyd, G.R.; Dewis, K.M.; Korshin, G.V.; Reiber, S.H.; Schock, M.R.; Sandvig, A.M.; & Giani, R., 2008. Effects of Changing Disinfectants on Lead and Copper Release. *Journal AWWA*, 100(11):75-87.	Literature review identifying key issues for predicting the effects of disinfectant change on lead and copper corrosion and release into drinking water.
Edwards, M. & Triantafyllidou, S., 2007. Chloride-to-sulfate Mass Ratio and Lead Leaching to Water. *Journal AWWA*, 99(7):96-109.	Study demonstrated the importance of CSMR in lead leaching.
Lytle, D.A. & Schock, M.R., 2005. Formation of Pb (IV) Oxides in Chlorinated Water. *Journal AWWA*, 97(11):102-114.	Study demonstrated the impact of ORP on lead mineral formation and solubility and how this was impacted by changes in water quality.
Lytle, D.A., 2007. Accumulation of Contaminants in the Distribution Systems. Cincinnati, Ohio: USEPA ORD/OGWDW Workshop on Inorganic Contaminant Issues.	Presentation describes USEPA research findings of types of contaminants found in DS accumulations, including corrosion by-products.

Reference	Description
Lytle, D.A. & Schock, M.R., 2008. *The Relationship Between Redox Stability, and Corrosion and Metal Release.* Cincinnati, Ohio: USEPA ORD, NRMRL, WSWRD, TTEB. Presented at AWWA WQTC.	Presentation describes USEPA research, which showed that variations in redox potential increased corrosion and metals release.
Nguyen, C.K.; Stone, K.R.; & Edwards, M.A., 2011. Chloride-to-Sulfate Mass Ratio: Practical Studies in Galvanic Corrosion of Lead Solder. *Journal AWWA,* 103(1):81-92.	Study demonstrated the importance of CSMR in lead leaching.
Singley, J.E., 1981. The Search for a Corrosion Index. *Journal AWWA,* 73(11):578-582.	Article describes and compares various commonly used corrosion indices.
USEPA, 1988a, 1991, 2004. Lead and Copper Rule & Guidance.	USEPA regulations and related guidance to control lead and copper at taps; corrosion control optimization.
USEPA, 2003. Revised Guidance Manual for Selecting Lead and Copper Control Strategies EPA-816-R-03-001.	Manual provides flowcharts and guidelines for selection of lead and copper treatment based on source water quality.
USEPA, 2007. Simultaneous Compliance Guidance Manual for the Long Term 2 and Stage 2 DBP Rules. EPA 815-R-07-017.	Manual provides integrated guidance and case studies to help utilities make best choices to comply with Stage 2 disinfection/DBP (D/DBP) and Long Term 2 Enhanced Surface Water Treatment Rule. Includes emphasis on potential for corrosion impacts relative to changes to disinfectants, etc.
USEPA, 2016. Optimal Corrosion Control Treatment Evaluation Technical Recommendations for Primacy Agencies and Public Water Systems. EPA 816-B-16-003.	Provides technical recommendations for determining the most appropriate treatment for controlling lead and copper and complying with LCR requirements.

Status

The status of internal corrosion control practices, as related to achieving a desired level of distribution system performance, may be assessed by reviewing the following self-assessment questions. The self-assessment team is not limited to these questions and may consider discussing additional topics related to internal corrosion control that may assist in identifying and addressing performance-limiting factors.

- Does the system meet regulatory lead and copper action levels?
  - ▲ For systems not located in the United States, consider whether the system meets all applicable regulatory requirements for lead and copper or associated corrosion-related parameters.
- Does the utility have a corrosion testing strategy (such as coupons, electronic detection, water quality monitoring) conducted routinely at locations throughout the system?

- Is the break site examined for tuberculation, pitting, holes, or scaling?
- Does the utility have a program established to address potential impacts on corrosion when changing source water and/or treatment conditions?

Action

If areas of the utility's distribution system internal corrosion control practices are considered to have a status of Partially Optimized or Not Optimized, and this performance improvement variable has been prioritized and selected for improvement, the following steps are recommended to be considered in the development of an action plan to improve performance in this area.

- If the system does not meet regulatory requirements for lead and copper or other corrosion-related parameters, develop a plan to help enable the utility to attain compliance.
    - ▲ This action may require the utility to determine the root causes of the noncompliance, which may be attributable to a variety of factors. Similarly, the plan to attain compliance may incorporate several factors including those related to source water, water chemistry and treatment conditions, as well as physical infrastructure.
- If the utility has not established a corrosion testing program, consider developing a program that enables the evaluation of corrosion and corrosion-related parameters, on a routine basis, throughout the system. At the most basic level, this may include monitoring for the key water quality parameters described earlier in this section. More advanced programs may also incorporate additional elements, such as electronic corrosion detection.
- If pipe condition is not currently examined for evidence of corrosion during main breaks and other opportunistic events, consider developing procedures that allow staff to inspect pipe condition whenever the pipe is exposed.
- Utilities without a written strategy for evaluating the potential impact of source water and/or treatment changes on corrosion are encouraged to develop a strategy and procedures to perform such an evaluation, incorporating the key components described earlier in this section.
    - ▲ Water quality parameters to consider addressing in such a policy include alkalinity/pH, disinfectant residual, chloride, sulfate, and the use of orthophosphate corrosion inhibitors. The potential impact of changes to these parameters includes:
        - Significant changes in alkalinity/pH may result in red water or black water episodes.

- Changes in disinfectant type or concentration may result in changes in DBP concentration, metal (i.e., lead and copper) corrosion, and/or release, and microbial changes.

- Significant shifts in the chloride to sulfate ratio may significantly impact lead and/or copper corrosion control.

- Changing from the use of orthophosphate to a treatment regime without orthophosphate may seriously disrupt the protective coating provided by the orthophosphate treatment. Once established, an orthophosphate scale must be maintained. Discontinuance of orthophosphate treatment can seriously disrupt the protective coating provided by orthophosphate treatment and lead to potentially significant lead and/or copper release and corrosion occurrence.

▲ Development of such a policy should also take into account the pipe materials present in the distribution system, with particular attention to unlined cast iron mains or areas with a history of manganese accumulation.

Performance-Limiting Factor Summary

Table 3-19 summarizes factors related to internal corrosion control that may be limiting optimized performance of the distribution system. Check whether these factors are Optimized and Documented, Partially Optimized, or Not Optimized. Factors identified as Partially Optimized or Not Optimized will be prioritized in chapter 7, Identification and Prioritization of Performance-Limiting Factors, where an action plan can be developed and implemented that allows for optimization of parameters selected for improvement.

Table 3-19. Internal corrosion control

Self-assessment category	Questions for gauging optimization status	Optimized and documented	Partially optimized	Not documented	Comments
		Response			
Lead and Copper Action Levels	Does the system meet regulatory lead and copper action levels?				
Corrosion Monitoring	Does the utility have a corrosion testing strategy (such as coupons, electronic detection, water quality monitoring) conducted routinely at locations throughout the system?				
Main Break or Leak Repair Testing	Is the break site examined for tuberculation, pitting, holes, or scaling?				
Change of Treatment and Source Water	Does the utility have a program established to address potential impacts on corrosion when changing source water and/or treatment conditions?				

NITRIFICATION

Understanding

Many utilities working to comply with DBP limits have successfully done so using chloramine as a secondary disinfectant. More utilities are considering conversion to chloramine because of increasingly strict DBP limits (40 CFR, USEPA 2007). Additionally, given similar treated water conditions, chloramination can be more effective than free chlorination at maintaining disinfectant residuals at locations with longer hydraulic retention times. Although chloramine is effective in minimizing TTHM and HAA5 formation in the distribution system, chloraminated systems can be prone to nitrification-related accelerated disinfectant residual loss in high water-age areas; therefore, it is important for chloraminated systems to frequently monitor for water quality parameters associated with nitrification, throughout the distribution system, and to develop procedures that enable distribution system staff to appropriately respond to potential nitrification events.

Additional references related to nitrification can be found in Table 3-20. The self-assessment questions for this Performance Improvement Variable are included in Table 3-21. This section of the self-assessment should be completed by all systems that use chloramine as a residual disinfectant, in any portion of the distribution system. Systems that use free chlorine only may address the questions in this portion of the self-assessment with a response of "Not Applicable."

Components of an Optimized System for Nitrification Management

Nitrification is a microbial process by which reduced nitrogen compounds (primarily ammonia) are sequentially oxidized to nitrite and nitrate. Ammonia is present in drinking water naturally or by ammonia addition during secondary disinfection to form chloramines. Free ammonia is metabolized to nitrite by ammonia oxidizing bacteria such as *Nitrosomonas*. Nitrite oxidizing bacteria such as *Nitrobacter* then convert nitrite to nitrate. Conditions that can lead to nitrification occurrence in drinking water include: overdosing of ammonia above the required level to form chloramine (presence of significant free ammonia), low chlorine residual, warm temperature (> 15°C), high water age, and the presence of conditions favoring biofilm formation. Left unchecked, nitrification produces increased nitrite, which can react with chloramine and cause rapid decay of chloramine and release of ammonia, promoting further nitrification. This cycling nature of the reaction between nitrification and chloramine can result in a rapid deterioration of water quality. The rapid decay of chloramine and loss of disinfectant residual then leads to increased HPC and coliform growth. Nitrification also results in reduced pH, alkalinity, and dissolved oxygen levels, which have the potential to increase corrosion (Zhang et al. 2009).

Nitrification control involves optimization of the chloramine treatment process as well as distribution system operation and maintenance. The key to nitrification control is to prevent its occurrence, well before water quality significantly deteriorates. Optimizing chloramine treatment by closely controlling the chlorine to ammonia ratio and decreasing water age by flushing and storage tank management are some of the most commonly used nitrification control strategies. When possible, utilities should reconfigure/redesign distribution systems to reduce dead ends and minimize water age.

Chloramine treatment should be optimized at the application point to minimize the potential for nitrification to occur. For example, a 4.5:1-5.0:1 chlorine to ammonia-N ratio is theoretically optimal to provide stable monochloramine residuals while minimizing the presence of excess free ammonia, although the exact ratio required to optimize monochloramine formation may vary from system to system (AWWA 2013). As the chlorine to ammonia-N ratio decreases, more free ammonia will be present in the finished water. As the chlorine to ammonia ratio increases above the optimal level, dichloramine formation can occur, which may lead to potential taste and odor issues in the distribution systems. At chlorine to ammonia ratios approaching that required for breakpoint chlorination (the point at which free chlorine is formed), the resulting trichloramine formation can lead to the presence of strong tastes and odors. Chloraminating systems often provide a chloramine dose that is adequate to result in a minimum total chlorine residual of 0.5-1.0 mg/L at system extremities. Chloramination is also pH dependent, with higher system pH (> 9.0) correlated with improved chloramine stability and reduced nitrification occurrence (AWWA 2013c).

The distribution system, including storage facilities, should be designed and operated to minimize water age. Distribution lines should be sized and configured (looped or directionally valved) to reduce dead ends, and storage should be designed, operated, and maintained to maximize storage cycling *and* mixing.

Seasonal switches from chloramine to free chlorine (often referred to by utilities as a free chlorine "burn") have been reported as helpful to prevent or reduce nitrification in distribution systems. Breakpoint chlorination of storage facilities or mains has been beneficial in instances of increased nitrification. Care must be taken when performing residual changeovers to inform consumers (such as those with special-use needs such as kidney dialysis, aquatic life keepers, or to prepare for aesthetic changes). Regulatory agencies typically require approval for disinfectant residual changes (AWWA/EES 2002b, AWWA 2013c) because such changeovers can result in increased DBP levels. It is necessary to properly control any discharges to the environment to protect aquatic life from chloramines. Utilities should also be aware that lead concentration levels may also be affected by changeovers between free chlorine and chloramine if the redox potential changes significantly (USEPA 2007). Partnership for Safe Water subscribers that use this practice may access chapter 2 for additional information regarding how changes in disinfectant are captured in the program's disinfectant residual data collection software.

To assess distribution system water quality with respect to nitrification, several key nitrification indicator parameters should be monitored regularly (AWWA 2012c, 2013c). These parameters include:

- Disinfectant residual
- Monochloramine
- Free and total ammonia
- Nitrite
- Heterotrophic bacteria
- pH

Systems should be aware that monitoring for free chlorine, using a DPD method, in the presence of monochloramine, can result in artificially high readings, because monochloramine is a positive interference with the free chlorine DPD chemistry.

Sampling should be conducted at the treatment plant, system entry points, and at locations where water age may be high such as:

- Storage facility outlets
- Pressure zone boundaries

- Large mains serving sparsely populated areas
- Dead ends

Utilities should set nitrification action levels (AWWA 2013c) and have appropriate response plans developed that are based on these action levels. Substantial changes in combinations of key nitrification parameters (such as rapid loss of chloramine residual, free ammonia increases, increased nitrite, increased HPC, and decreased pH) are signs of nitrification. Actions taken in response to nitrification need to be quick and appropriate for the situation, ranging from verification of source treatment, increased water turnover in affected areas, or breakpoint chlorination (AWWA/EES 2002b, AWWA 2013c). Utilities receiving chloraminated water from a wholesaler should design and install equipment to monitor for disinfectant residual concentration and nitrification-related parameters. If chloramine levels in the water received are lower than optimal, it may be necessary to boost chlorine and ammonia in order to optimize the concentration of chloramine delivered to the system or work with the water wholesaler to improve the quality of water that is delivered.

Status

The status of nitrification control practices, as related to achieving a desired level of distribution system performance, may be assessed by reviewing the following self-assessment questions. The self-assessment team is not limited to these questions and may consider discussing additional topics related to nitrification control that may assist in identifying and addressing performance-limiting factors.

- Are free ammonia, nitrite, and heterotrophic plate count (HPC) tested routinely? Are action levels established that are based on test results?
- Is the total chlorine residual maintained at a concentration greater than 0.50 mg/L? Are storage tanks monitored, particularly in areas lacking circulation? Are zone boundaries and dead ends monitored for nitrification?

Action

If areas of the utility's distribution system nitrification control practices are considered to have a status of Partially Optimized or Not Optimized, and this performance improvement variable has been prioritized and selected for improvement, the following steps are recommended to be considered in the development of an action plan to improve performance in this area.

- If chloramine concentrations and nitrification-related parameters (monochloramine, free ammonia, nitrite, HPC) are not monitored routinely throughout the distribution system, develop a plan to monitor for these

Table 3-20. References for effective nitrification control

Reference	Description
AWWA/EES (Economic and Engineering Services), 2002b. *Nitrification.*	USEPA Total Coliform Rule issue paper describes nitrification phenomenon and cites many key references (research and utility case studies). Explanations of multifactorial causes and effective monitoring and control programs.
AWWA, 2012. *Standard Methods for Examination of Water and Wastewater.* 22nd ed. E.W. Rice, R.B. Baird, A.D. Eaton, L.S. Clesceri, editors. ISBN 978-0-87553-013-0	Methods for nitrification-related WQP analysis (e.g., free and total chlorine, monochloramine, free and total ammonia, nitrite, HPC) are described in detail along with interferences.
AWWA, 2013. M56, *Nitrification Prevention and Control in Drinking Water, Second Edition.* Denver, Colo.: AWWA.	Manual provides information on the occurrence and microbiology of nitrification in water and provides approaches for nitrification prevention or detection and mitigation.
CFR Title 40—Chapter 1, Subchapter D.	USEPA federal drinking water regulations specify samples required methods/location of sampling, analytical methods for disinfectant and disinfection by-products, inorganic chemicals, and microbials.
USEPA, 2007. Simultaneous Compliance Guidance Manual for the Long Term 2 and Stage 2 DBP Rules. EPA 815-R-07-017.	Manual provides integrated guidance and case studies to help utilities make best choices to comply with Stage 2 DDBP and Long Term 2 Enhanced Surface Water Treatment Rule rules. Includes emphasis on potential for nitrification-related impacts when using chloramine and cites corrosion impacts relative to nitrification and chloramine.
Zhang, Y.; Love, N.; Edwards, M., 2009. Nitrification in Drinking Water Systems. *Critical Reviews in Environmental Science and Technology,* 39(3):153-208.	Review of research on nitrification as it relates to the ammonia levels and unique environments present in potable water distribution systems. Factors affecting nitrification occurrence, nitrification impacts on water quality and corrosion, and nitrification monitoring and control methods are emphasized.

parameters, particularly at sites that may exhibit an increased risk for nitrification. As stated previously in this section, nitrification is best controlled through early detection and prevention, well before water quality significantly deteriorates.

▲ If testing for nitrification-related parameters is a new procedure for utility staff, ensure that staff is properly trained on all procedures and have the ability to produce accurate test results.

- If total chlorine is not maintained at adequate levels throughout the system, develop a strategy to enable the system to meet the Partnership's performance goals (or local regulatory requirements, if they are higher than the Partnership's goals) for distribution system disinfectant residuals.

  ▲ Consider prioritizing areas at high risk for residual loss when developing such a strategy.

  ▲ Utilities that purchase water from a wholesaler may benefit from communicating with the wholesaler to further discuss water chemistry or disinfectant residual issues.

Performance-Limiting Factor Summary

Table 3-21 summarizes factors related to nitrification control that may be limiting optimized performance of the distribution system. Check whether these factors are Optimized and Documented, Partially Optimized, or Not Optimized. Factors identified as Partially Optimized or Not Optimized will be prioritized in chapter 7, Identification and Prioritization of Performance-Limiting Factors, where an action plan can be developed and implemented that allows for optimization of parameters selected for improvement.

Table 3-21. Nitrification control (systems that use chloramine disinfection)

Self-assessment category	Questions for gauging optimization status	Optimized and documented	Partially optimized	Not optimized	Comments
Nitrification Detection and Control	Are free ammonia, nitrite, and heterotrophic plate count (HPC) tested routinely? Are action levels established that are based on test results?				Assess utility action level goals for these parameters.
Nitrification Control	Is the total chlorine residual maintained at a concentration greater than 0.50 mg/L? Are storage tanks monitored, particularly in areas lacking circulation? Are zone boundaries and dead ends monitored for nitrification?				Use disinfectant residual performance spreadsheet to assist with this evaluation.

PIPELINE INSTALLATION, REHABILITATION, AND REPLACEMENT

Understanding

Pipeline installation, rehabilitation, and replacement practices have a long-term effect on water distribution system infrastructure condition and performance. By following industry standards (such as Ten State Standards and ANSI/AWWA C600-10, C605-13) for the proper design and installation of pipelines, utilities can maximize asset life, minimize leaks and breaks, and deliver the desired water quantity and quality to the customers they serve. The information in this section focuses on helping a utility prepare and carry out a comprehensive pipeline renewal program that is composed of a combination of rehabilitation and replacement projects.

Additional references related to pipeline installation, rehabilitation, and replacement can be found in Table 3-22. The self-assessment questions for this Performance Improvement Variable are included in Table 3-23.

Components of an Optimized System for Pipeline Rehabilitation and Replacement

Pipeline Inventory

To help a utility manage a pipeline rehabilitation and replacement program, the utility should have an inventory of pipelines that includes both physical and location attributes. Very small systems may be able to manage this information effectively without an electronic database, but it is recommended that most utilities manage this information inventory electronically so that it can be easily sorted, queried, and updated. Commercially available geographic information system (GIS) software packages are most suitable for linear assets, such as pipelines, and can aid in the collection, storage, retrieval, sorting, and analysis of pipeline physical and location data. At a minimum, physical data in the inventory should include pipe size, material, class, lining, and the year of installation. Joint type, manufacturer, soil conditions, external coatings and protection, and loading conditions are also useful physical data to record. Location data should include the street name, municipality, and pressure zone. The effective management of physical asset and location information for pipelines can aid in a utility's ability to locate and repair mains and plan for pipeline rehabilitation or replacement projects.

Performance Assessment and Tracking

Most pipeline performance information is obtained from a combination of reactive data collection (customer complaints, breaks, and leaks) and routine monitoring (water quality, flow, and pressure). Main breaks and leaks are usually considered to be the most significant and complex dataset (Deb et al. 2002, Grigg

2004). Because deterioration of pipe condition is often the cause of unexpected performance failure, some utilities use additional methods to assess pipe condition. These methods include, but are not limited to, acoustic leak detection, sensors on or within transmission mains, soil corrosiveness testing, and pipe coupon testing. Nondestructive inspection of pipelines is also common practice for larger mains, in which the consequence of failure is high. In reality, it is often difficult to assess all components, so condition assessment methods must also focus on systems by aggregating information about system components (Grigg 2004).

Planning and Prioritization

The utility should have a written and systematic pipeline rehabilitation and replacement program. Two key challenges utilities face are determining the appropriate annual funding (macrolevel planning) and selecting the specific main segments to be rehabilitated or replaced (microlevel planning). Utilities should invest capital in rehabilitation and replacement projects that have been assessed as having high likelihood of failure and the most serious consequences.

Grigg (2004) provides several good examples of existing utility programs and describes using the KANEW software and Nessie curves for macrolevel planning. For microlevel planning, Deb (2002), Grigg (2004), and AWWA M28 (AWWA 2014b) describe several reactive and predictive approaches for prioritizing rehabilitation and replacement of mains. Additionally, some models are commercially available. AWWA's Buried No Longer forecasting tool is an example of a utility-accessible tool for modeling pipe replacement and rehabilitation requirements. These approaches include scoring methods (for example, point systems), economic analysis (such as break-even or cost-benefit analysis), failure probability and regression analyses, and mechanistic models to predict pipe failure based on loading and condition.

Most methods have significant data needs and provide results that are not absolute and must be applied with some subjective judgment from utility management. Rehabilitation and replacement programs should consider options for maintenance and repair (proactive or reactive), cathodic protection (for metal pipe only), relining (nonstructural or structural), or replacement (open cut or trenchless) as described by Grigg (2004). The trenchless replacement of water mains option is also described well by Chapman (2007).

Status

The status of pipeline installation, replacement, and rehabilitation programs, as related to achieving a desired level of distribution system performance may be assessed by reviewing the following self-assessment questions. The self-assessment team is not limited to these questions and may consider discussing additional

Table 3-22. References for pipeline installation, rehabilitation, and replacement programs

Reference	Description
AWWA, 2010. ANSI/AWWA C600-10, Installation of Ductile Iron Water Mains and Their Appurtenances. Denver, Colo.: AWWA.	Provides detailed guidelines for ductile iron pipe handling, installation, and verification.
AWWA, 2011. ANSI/AWWA C602-11, Cement-Mortar Lining of Water Pipelines in Place—4 in. (100 mm) and Larger. Denver, Colo.: AWWA.	Provides detailed guidelines for preparation and application of cement mortar lining of water pipelines.
AWWA, 2013. ANSI/AWWA C605-13, Underground Installation of Polyvinyl Chloride (PVC) and Molecularly Oriented Polyvinyl Chloride (PVCO) Pressure Pipe and Fittings. Denver, Colo.: AWWA.	Provides detailed guidelines for PVC pipe handling, installation, and verification.
AWWA, 2015. ANSI/AWWA G200-15, Distribution Systems Operation and Management. Denver, Colo.: AWWA.	Establishes a standard with minimum requirements for pipeline rehabilitation and replacement programs.
AWWA, 2005. *Water Distribution System Assessment Workbook*. Smith, C. editor. Denver, Colo.: AWWA.	Provides checklists and additional references for evaluating water pipeline renewal programs.
AWWA, 2014. M28, *Rehabilitation of Water Mains, Third Edition*. Denver, Colo.: AWWA.	Provides background information and guidance for selecting water main rehabilitation methods including cleaning, lining, and trenchless replacement.
Chapman, D.N.; Ng, P.C.F.; & Karri, R., 2007. Research Needs for On-line Pipeline Replacement Techniques. *Tunnelling and Underground Space Technology*, 22(5-6):503-514.	Describes various methods of trenchless replacement of water mains.
Deb, A.K.; Grablutz, F.; Hasit, Y.; Snyder, J.; Loganathan, G.; & Agbenowsi, N., 2002. AwwaRF #459. Prioritizing Water Main Replacement and Rehabilitation. Denver, Colo.: AWWA Research Foundation.	Provides methods for analyzing main break data and prioritizing pipeline renewal.
Great Lakes Upper Mississippi River Board of State and Provincial Public Health and Environmental Managers (GLUMRB), 2007. Recommended Standards for Water Works. Albany, N.Y.: Health Research Inc., Health Education Services Division.	Establishes standards for design and construction of water distribution systems, including materials selection and proximity to sewers.
Grigg, N.S., 2004. AwwaRF #91025F. Assessment and Renewal of Water Distribution Systems. Denver, Colo.: AWWA Research Foundation.	Provides extensive compilation of techniques and references for planning water distribution system renewal programs.

topics related to pipeline installation, replacement, and rehabilitation programs that may assist in identifying and addressing performance-limiting factors.

- Are all pipelines installed and disinfected as required by applicable regulations? Do procedures follow AWWA/ANSI Standard C600-620 and C651 or others, as appropriate?

- Is there a formal process to prioritize main replacements? Is the rate of replacement and rehabilitation adequate to reduce the amount of unlined metal pipe in the system?

- Is the renewal rate adequate to reduce the pipe mileage that has the highest risk of failure?

Action

If areas of the utility's pipeline installation, replacement, and rehabilitation programs are considered to have a status of Partially Optimized or Not Optimized, and this performance improvement variable has been prioritized and selected for improvement, the following steps are recommended to be considered in the development of an action plan to improve performance in this area.

- If pipelines are not installed and disinfected as required, establish procedures that enable the utility to comply with all applicable pipeline installation and disinfection requirements, as referenced previously in this section. Ensure that staff is adequately trained to properly perform all required procedures.

- If a formal process does not exist, consider the benefits of developing a formal process to prioritize required main replacement projects. It is recommended to consider developing a process that allows the utility to prioritize and budget for pipeline rehabilitation and replacement projects to be completed at a renewal rate that is adequate to address the pipeline in the system at highest risk of degradation and/or failure.

    ▲ A variety of industry resources exist to support these efforts, including commercially available software programs. It is recommended that utilities access these resources to assist in the development of this process.

    ▲ For utilities that do not already have the information, it is recommended to include resources for the completion of a system inventory and condition assessment/tracking.

Performance-Limiting Factor Summary

Table 3-23 summarizes factors related to pipeline installation, replacement, and repair that may be limiting optimized performance of the distribution system. Check whether these factors are Optimized and Documented, Partially Optimized, or Not Optimized. Factors identified as Partially Optimized or Not Optimized will be prioritized in chapter 7, Identification and Prioritization of Performance-Limiting Factors, where an action plan can be developed and implemented that allows for optimization of parameters selected for improvement.

Table 3-23. Pipeline installation, replacement, and rehabilitation

Self-assessment category	Questions for gauging optimization status	Optimized and documented	Partially optimized	Not optimized	Comments
Pipeline Installation	Are all pipelines installed and disinfected as required by applicable regulations? Do procedures follow AWWA/ANSI Standards C600-620 and C651 or others, as appropriate?				
Main Replacement and Rehabilitation	Is there a formal process to prioritize main replacements? Is the rate of replacement and rehabilitation adequate to reduce the amount of unlined metal pipe in the system?				
Pipeline Renewal Rate	Is the renewal rate adequate to reduce the pipe mileage that has the highest risk of failure?				

POST-PRECIPITATION, INORGANIC ACCUMULATION

Understanding

Post-precipitation refers to substances that can precipitate after treatment processes. Common examples of solids that may precipitate out in the distribution system, along with their sources, are listed here.

- Aluminum or iron from coagulants
- Zinc or phosphates from corrosion inhibitors

- Carbonate scales and particulates from pH or alkalinity adjustment chemicals or source water components
- Manganese from source water or treatment residuals
- Arsenic from source water

Post-precipitation can pose challenges to achieving distribution system optimization. Some of the accumulated substances can exert a disinfectant demand, and all can contribute to scale and sediment that can harbor microorganisms. Inorganic substance accumulation refers to the chemical and physical process that can allow for increasing levels of metals, metalloids, and other contaminants within the scales and sediments of the distribution system.

As discussed in Friedman et al. (2010), the particular concern with inorganic substance accumulation is not the physical presence of substances in on the distribution system piping materials, but rather the potential for their remobilization into the water. The release of collected contaminants in concentrated amounts can result in levels at customer taps that are of public health concern (Schock et al. 2008).

Additional references pertaining to post-precipitation and inorganic accumulation can be found in Table 3-24. The self-assessment questions for this Performance Improvement Variable are included in Table 3-25.

Components of an Optimized System for Postprecipitation Control

Chemical precipitates (potentially composed of unregulated or regulated inorganic substances) may form in the distribution system depending on water quality conditions. Chemical precipitates may deposit onto and coat the surfaces of piping, reservoirs, and plumbing systems. Chelating and sequestering agents (such as polyphosphate and silica) can form complexes with various inorganic elements (such as iron, copper, or lead), thus influencing the potential for precipitation and deposition within the distribution system (Friedman et al. 2010). Other chemical and physical processes such as adsorption, absorption, and corrosion can also cause substantial amounts of chemical solids to accumulate in distribution systems (Schock 2005, USEPA 2006e, Benjamin 2014).

Both turbidity and color may serve as potential indicators for chemical precipitates or inorganic particulates, but analytical methods for these parameters may be subject to interferences from organics, microbiological components, or dissolved inorganic substances. Therefore, turbidity or color measurements only correlate indirectly with the water's chemical or biological characteristics (AWWA 2012c). In light of this, Friedman (2010) describes a three-step program to assess and control accumulation of inorganic substances.

Step 1—Assess Existing Conditions and Vulnerability

Determine the prevalence of deposits on existing infrastructure by assessing the adequacy of flushing programs and evaluating treated water quality conditions.

Step 2—Address Existing Deposits

Identify, prioritize, and replace or rehabilitate deteriorated pipes. Focus immediate resources on removing loosely adhered solids from the distribution system and implementing measures to stabilize adhered solids to the greatest extent possible. These measures essentially represent "release management" techniques. Mobile deposits should be removed through high velocity unidirectional flushing. Conventional flushing is not recommended since contaminants can be stirred up and delivered to the consumer. Adhered deposits are removed through pipe rehabilitation or replacement, and stabilized by improved water treatment.

Step 3—Reduce Contaminant and Solids Loading

Reduce the potential for continued contaminant loading to the distribution system by implementing treatment to remove trace contaminants, avoiding "sidestream" treatment or blending to meet MCLs, and evaluating chemical sequestration if removal is not possible. Iron and manganese serve as "substrate solids" for the accumulation of regulated heavy metals. Removing these substances (although not usually needed) can potentially result in benefits that go beyond aesthetic improvements.

Status

The status of post-precipitation control practices, as related to achieving a desired level of distribution system performance may be assessed by reviewing the following self-assessment questions. The self-assessment team is not limited to these questions and may consider discussing additional topics related to post-precipitation control that may assist in identifying and addressing performance-limiting factors.

- Are areas of low disinfectant residual investigated for precipitation?
- Are treatment plant practices optimized to reduce precipitation potential?
- Are areas of precipitation accumulation cleaned or flushed at least annually (storage tanks, low water-use areas, dead-end mains)?

Action

If areas of the utility's post-precipitation control practices are considered to have a status of Partially Optimized or Not Optimized, and this performance

Table 3-24. References for control of post-precipitation inorganics accumulation

Reference	Description
AWWA, 2012. *Standard Methods for Examination of Water and Wastewater.* 22nd ed. E.W. Rice, R.B. Baird, A.D. Eaton, and L.S. Clesceri, editors. ISBN 978-0-87553-013-0.	Methods for analysis of chemical, physical, and microbial parameters are described in detail along with interferences and calibration related issues, etc.
AWWA/EES (Economic and Engineering Services), 2002. *Effects of Water Age on Distribution System Water Quality.*	Soft waters, low pH, alkalinity, high Cl- and/or SO4, high DO, low buffer intensity, high conductivity, high Cl residual (aggressive waters) may lead to corrosion or scaling-related accumulations.
Benjamin, M.M., 2014. *Water Chemistry*, 2nd ed. Long Grove, Ill: Waveland Press Inc.	Textbook explains chemical precipitation and related phenomena that occur in water.
Friedman, M.; Martel, K.; Hill, A.; Holt, D.; Smith, S.; Ta, T.; Sherwin, C.; Hiltebrand, D.; Pommerenk, P.; Hinedi, Z.; & Camper, A., 2003. AwwaRF #2606. Establishing Site-Specific Flushing Velocities. Denver, Colo.: AWWA Research Foundation.	Research explains, documents flushing velocities needed to minimize solids accumulations in distribution systems.
Friedman, M.J.; Hill, A.; Korshin, G.; Valentine, R.; & Reiber, S., 2010. WRF #3118. Assessment of Inorganics Accumulation in Drinking Water System Scales and Sediments. Denver, Colo.: Water Research Foundation.	Extensive field study of chemical components of distribution system scales and sediments. Detailed discussion on accumulation and release mechanisms, as well as control strategies.
Lytle, D.A., 2007. Accumulation of Contaminants in the Distribution Systems. Cincinnati, Ohio: USEPA ORD/OGWDW Workshop on Inorganic Contaminant Issues..	Presentation describes USEPA research findings of types of contaminants found in DS accumulations including corrosion byproducts.
Schock, M.R., 2005. Distribution Systems and Reservoirs and Reactors for Inorganic Contaminants (Chapter 6). In: *Distribution System Water Quality Challenges in the 21st Century.* Denver, Colo.: AWWA.	Textbook chapter describes variety of inorganic contaminant reactions that can occur in distribution systems. Numerous citations and examples are provided from USEPA and other research.
USEPA, 2006e. Inorganic Contaminant Accumulation in Potable Water Distribution Systems. Washington D.C.: USEPA. Total Coliform Rule Issue Paper.	USEPA Total Coliform Rule issue paper summarizes USEPA regulations of inorganic chemicals, some of which can be incorporated into precipitated solids via adsorption, absorption, or co-precipitation.
USEPA, 2008b. National Secondary Drinking Water Regulations CFR 40 Part 143. e-CFR.	USEPA regulations for aesthetics-based parameters that can contribute to chemical precipitation and accumulation.

improvement variable has been prioritized and selected for improvement, the following steps are recommended to be considered in the development of an action plan to improve performance in this area.

- If areas of low disinfectant residual are not routinely investigated for precipitation, consider developing SOPs that incorporate this into the utility's response to low disinfectant residual occurrence.

- Consider the potential for improving treatment plant practices to reduce the potential for metals accumulation and precipitation in the distribution system.
  - ▲ A variety of treatment techniques may be evaluated for this, including chemical sequestration and removal methods, such as coagulation and filtration, to reduce the concentrations of metals entering the distribution system.

- If areas of precipitation accumulation are not cleaned or flushed on an annual basis, consider developing SOPs that allow for this practice.
  - ▲ Areas in which precipitation frequently occur include storage tanks, low water-use areas, and dead-end mains, although utility staff should identify any additional areas in which precipitation occurs.
  - ▲ Techniques for the removal of precipitates include flushing as well as physical pipe cleaning methods.

Performance-Limiting Factor Summary

Table 3-25 summarizes factors related to post-precipitation control that may be limiting optimized performance of the distribution system. Check whether these factors are Optimized and Documented, Partially Optimized, or Not Optimized. Factors identified as Partially Optimized or Not Optimized will be prioritized in chapter 7, Identification and Prioritization of Performance-Limiting Factors, where an action plan can be developed and implemented that allows for optimization of parameters selected for improvement.

Table 3-25. Post-precipitation control

Self-assessment category	Questions for gauging optimization status	Optimization status			Comments
		Optimized and documented	Partially optimized	Not optimized	
Precipitation Recognition	Are areas of low disinfectant residual investigated for precipitation?				
Precipitation Control	Are treatment plant practices optimized to reduce precipitation potential?				
Precipitation Remediation	Are areas of precipitation accumulation cleaned or flushed at least annually (storage tanks, low water-use areas, dead-end mains)?				Evaluate the plan to reduce or eliminate continued precipitation.

SECURITY AND ONLINE MONITORING

Understanding

Real-time monitoring can serve an important role to alert utilities to events causing distribution system water quality changes (Hall et al. 2004; USEPA 2004, 2005). Online monitors need to provide real-time data and have the appropriate reliability, sensitivity, and specificity to assist utility staff in rapidly detecting and responding to the presence of a variety of potential distribution system contaminants. Ideally, distribution system water quality monitors should provide accurate information, required minimal maintenance, and provide the necessary degree of automation to allow for remote data capture and response.

Additional references related to security and online monitoring can be found in Table 3-27. The self-assessment questions for this Performance Improvement Variable are included in Table 3-28.

Components of an Optimized System for Security Management

An almost limitless number of compounds can potentially contaminate drinking water; therefore, monitoring with compound-specific sensors would still potentially leave a system vulnerable to contamination by those compounds that were not being actively monitored. To maintain proactive vigilance for distribution system water quality, optimized utilities monitor various water quality surrogates that can "cast a wide net" to capture changes in water quality that arise from various chemical or microbiological agents that are not detectable by human senses (Hall at al. 2004). Monitoring surrogate parameters allows for the detection of changes in water quality, irrespective of whether these changes are intentional (sabotage, vandalism) or accidental (for example, cross-connection backflow, intrusion, disruption of sediments, or biofilms). A list of several surrogate parameters and their application in distribution system water quality monitoring is provided in Table 3-29.

Examples of parameters that can be monitored continuously (online) and are often useful for detecting changes in water quality patterns include disinfectant residual, pH, turbidity, specific conductance, oxidation reduction potential (ORP), and total organic carbon (TOC). Pressure monitors can also be informative in indicating the physical and hydraulic characteristics of the system. Additionally, software systems are available commercially to help utilities establish the baseline water quality characteristics of the system and recognize the chemical changes that may be representative of different contaminant types, so that changes in overall water quality can be better detected and appropriately responded to. In addition to those detected by instrumentation, water quality changes detected by consumers should always be taken seriously as a potential warning of water quality integrity deterioration.

The USEPA has performed studies with water systems that provided guidance about the number of sensors used throughout the system versus levels of public health protection. The studies showed an improvement in public health protection with placement of up to 30 sensors at representative locations in the system, based on distribution size and complexity.

To judge on whether substantial changes in water quality have occurred, it is necessary to understand the normal or baseline water quality characteristics of the system. Monitoring water quality parameters regularly, at least at the major POE to the distribution system, helps to provide this information. However, it is also beneficial to establish a performance baseline at a variety of distribution system locations, particularly those that may be at a particular risk for water quality degradation. Additionally, on-line security-related monitoring guidelines include the following:

- Small systems (< 10,000 population)
  - ▲ At least one site (placed for maximum coverage) monitoring chlorine, pH, turbidity, temperature, conductivity, and pressure.
- Medium systems (10,000-100,000 population)
  - ▲ At least three sites monitoring chlorine residual, pH, turbidity, temperature, and conductivity; at least five pressure monitoring sites and one TOC monitoring point at the point of entry (such as the largest POE); and one within the system (for systems with population > 50,000).
- Large systems (> 100,000 population)
  - ▲ At least six sites monitoring chlorine, pH, turbidity, temperature, and conductivity, at least 10 pressure monitoring sites, and TOC monitoring of the largest POE plus two sites in the system.

Given the wide variety of sensor types and configuration possibilities, and the initial and operation and maintenance costs, utilities should investigate

Table 3-26. Surrogate parameters for detecting changes in water quality

Surrogate parameter	Application in distribution system water quality monitoring
Disinfectant residual	Distribution systems maintain residual concentrations (free or total chlorine, chlorine dioxide) for the disinfection of bacteria and viruses and the prevention of regrowth. Sudden drops in residual concentration may be indicative of chlorine demand from nitrogen or organic compounds.
TOC	Measures the total amount of organic carbon in the water, but does not identify specific organic compounds. More specialized instrumentation (gas chromatography/mass spectrometry, liquid chromatography/mass spectrometry) is required to identify specific organic constituents.
Conductivity	Measures the capacity of water to carry an electrical current. Conductivity is an indicator of the level of dissolved solids (salts) and can identify changes that may represent a change in overall water quality.
Turbidity	Indicator of suspended matter and microorganisms. Unexpected turbidity changes may be indicative of the presence of a contaminant (chemical or microbial).
Adenosine triphosphate and polymerase chain reaction (ATP and PCR)	Detection in drinking water may indicate the presence of biological activity.
Biosensors (toxicity)	The use of biosensors (such as luminescent tests for the presence of microbiological activity) can help to detect changes in microbiological activity in the distribution system.
pH	Indicator of hydrogen ion activity (acidity or basicity)of water. Carbon dioxide/bicarbonate/carbonate and ammonia/ammonium equilibria are pH dependent. Disinfection, metals solubility, and many other chemical and biochemical processes are pH dependent.
Pressure	Steady system pressures are desirable. Sudden changes may be indicative of breaks and leaks. Pressure transients may lead to intrusion events
Oxidation reduction potential	Changes in ORP may indicate the introduction of contaminants or other system conditions that affect disinfectant residual concentration levels.

all factors and options before selection and installation of online monitors. It is important to ensure that the selected instruments are appropriate for their monitoring application and installation environment. Correct installation, setup, and continuing maintenance requirements are also critical to their usefulness. Online monitors also need proper remote communication and data collection capabilities through automated computer systems, including appropriate alarm set points. These data then need to be regularly monitored and evaluated to allow for appropriate adjustments of the sensors over time.

The information generated by analyzers may be monitored using event detection software or by using automated alarm set points for the relevant monitoring parameters. Procedures then need to respond appropriately to alarm conditions. Advanced expertise and ongoing operator training should be provided to enable operators to use real-time trends. Providing access to archived data for effective routine and emergency response purposes is also recommended. It is important to consider any operational changes or changes in source water or distribution system blending when investigating possible causes of sudden water quality changes. Where continuous monitors are installed in distribution systems, established and updated baseline data should be included in routine optimization efforts and emergency response planning. During emergency events, water quality data may be correlated with related information, such as information from security monitoring systems, customer complaint surveillance systems, and public health surveillance channels (Pickard 2011).

Status

The status of security and online monitoring practices, as related to achieving a desired level of distribution system performance may be assessed by reviewing the following self-assessment questions. The self-assessment team is not limited to these questions and may consider discussing additional topics related to security and online monitoring that may assist in identifying and addressing performance-limiting factors.

- Have the items identified in the distribution system vulnerability assessment/analysis been implemented?

- Is there an emergency response plan? Are plan exercises conducted regularly? Are items identified during an exercise followed up?

- Are disinfectant residual records readily available for reference during an emergency?

Action

If areas of the utility's security and online monitoring practices are considered to have a status of Partially Optimized or Not Optimized, and this performance improvement variable has been prioritized and selected for improvement, the following steps are recommended to be considered in the development of an action plan to improve performance in this area.

- If items identified in the distribution system vulnerability assessment/analysis have not been implemented, develop a plan to implement actions to address these items. It may be helpful to prioritize the findings from the vulnerability assessment such that the utility is addressing the items with the greatest potential impact and urgency before addressing less significant items.

Table 3-27. References for security and online monitoring

Reference	Description
AWWA, 2010. ANSI/AWWA J100-10 (R13), Risk and Resilience Management of Water and Wastewater Systems (RAMCAP). Denver, Colo.: AWWA.	Sets requirements for risk and resilience analysis and management for the water sector and prescribes methods that can be used for addressing these requirements.
AWWA, 2001. M19, *Emergency Planning for Water Utilities, Fourth Edition*. Denver, Colo.: AWWA.	Offers guidance and tools for water and wastewater managers in preparing for emergencies.
Hall, J.; Zaffiro, A.D.; Marx, R.B.; Kefauver, P.C.; Krishnan, E.R.; Haught, R.C.; & Herrmann, J.G., 2007. On-line Water Quality Parameters as Indicators of Distribution System Contamination. Journal AWWA (99:1):66-77.	USEPA initiated program to investigate how changes in water quality parameters, which potentially indicate contamination, may be detected by real- or near real-time sensors.
Pickard, B.C.; Haas, A.J.; & Allegier, S.C., 2011. Optimizing Operational Reliability of the Cincinnati Contamination Warning System. *Journal AWWA*. 103(1):60-68.	Describes the operational reliability of a pilot contaminant warning system, based on analysis of specific performance metrics.
States, S., 2010. *Security and Emergency Planning for Water and Wastewater Utilities*. Denver, Colo.: AWWA.	Practical reference on security and emergency preparedness, planning, and implementation for water or wastewater utility managers and operators.
USEPA, 2004. Response Protocol Toolbox and Guidelines: Planning for and Responding to Drinking Water Contamination Threats and Incidents Module 1: Water Utilities Planning Guide. Interim final August 2004. Office of Ground Water and Drinking Water Security Division EPA 817-D-04-001 2003, EPA 817-D-04-001 2004.	Response guidelines are intended as a "field guide" for responding to contamination threats and can be developed in many different formats.
USEPA, 2005. WaterSentinel Online Water Quality Monitoring as an Indicator of Drinking Water Contamination. *Draft*, Version 1.0 EPA 817-D-05-002, EPA 817-D-05-001.	WaterSentinel serves as a demonstration project, or pilot, for designing and implementing an effective contamination warning system in a drinking water distribution system. This document describes the state-of-the-science of real-time, online water quality monitoring using conventional water quality parameters.

- If an emergency response plan has not been developed or is not exercised on a regular basis, consider developing a comprehensive emergency response plan, as well as a schedule for conducting emergency response exercises on a regular basis.

   ▲ AWWA's Standard J100—Risk and Resilience Management of Water and Wastewater Systems and M19—Emergency Planning for Water Utilities may be useful sources of information for completing this process.

- It is helpful for disinfectant residual records to be readily accessible to those who require them in an emergency situation. If this is not the case, consider taking steps to ensure the availability of disinfectant residual data, as well as that for other common water quality parameters, to those who require it to develop an appropriate emergency response.

Performance-Limiting Factor Summary

Table 3-28 summarizes factors related to security and online monitoring practices that may be limiting optimized performance of the distribution system. Check whether these factors are Optimized and Documented, Partially Optimized, or Not Optimized. Factors identified as Partially Optimized or Not Optimized will be prioritized in chapter 7, Identification and Prioritization of Performance-Limiting Factors, where an action plan can be developed and implemented that allows for optimization of parameters selected for improvement.

STORAGE FACILITY OPERATION AND MAINTENANCE

Understanding

Storage facilities have the potential to significantly impact water quality. Water quality can be affected if stored water becomes contaminated because of structural breaches in storage facilities, improper installation or operations and

Table 3-28. Security, online monitoring

Self-assessment category	Questions for gauging optimization status	Optimized and documented	Partially optimized	Not optimized	Comments
Vulnerability Assessment	Have the items identified in the distribution system vulnerability assessment/analysis been implemented?				
Emergency Response	Is there an emergency response plan? Are plan exercises conducted regularly? Are items identified during an exercise followed up?				Evaluate the frequency and effectiveness of exercises.
Disinfectant Residual Monitoring	Are disinfectant residual records readily available for reference during an emergency?				

maintenance (O&M) practices, or if storage is inadequately cycled or mixed. An active program to inspect storage tanks for structural defects, leaks, corrosion, or accumulation of sediment is an important part of a storage facility maintenance program. ANSI/AWWA Standards D102 (ANSI/AWWA D102-14) and D103 (ANSI/AWWA D103-09) address maintenance activities such as coatings and painting. ANSI/AWWA Standard D104 (ANSI/AWWA D104-11) covers cathodic protection of steel tanks. Storage tank mixing is affected by design *and* operational factors, which are described in many research publications (Kirmeyer et al. 1999, 2000; Grayman et al. 2000; Mahmood et al. 2005; Roberts et al. 2006). It is beneficial to monitor water quality in storage tanks, as described in various sections throughout this guide.

Additional references related to storage facility O&M can be found in Table 3-29. The self-assessment questions for this Performance Improvement Variable are included in Table 3-30.

Components of an Optimized System for Storage Facility O&M
Storage Facility Water Quality

The most common water quality problem experienced in finished water reservoirs is the loss of disinfectant residual. Kirmeyer (1999) recommends a five-step approach to establishing standard operating procedures to address storage facility water quality. These steps include: (1) understand the facility, (2) define the problems, (3) evaluate alternatives, (4) implement good management practices and monitor effectiveness, and (5) develop SOPs. Water quality monitoring in storage tanks is an important component to incorporate into these steps and can be beneficial in helping utilities to better define the problems that they may be experience.

In the *Guidance Manual for Maintaining Distribution System Water Quality*, Kirmeyer (2000) provides recommendations for operating storage facilities to improve water quality. Friedman (2005) provides an example of a Comprehensive Management Plan for Finished Water Storage Facilities that establishes clear guidelines and procedures for the management, operation, maintenance, inspections, planning, and design of finished water storage facilities.

Tank Capacity and Turnover

The utility should have written operating procedures that define the optimal water level changes in the storage facilities to ensure acceptable pressures and fire flows. These operating plans should also establish targeted turnover rates to minimize water age in finished water storage facilities and maintain an acceptable level of water quality throughout the storage tank. Tank turnover time can generally be estimated by comparing the average tank volume to the daily drawdown, although caution should be used that this is not an exact calculation. A wide variety of storage tank mixing systems are commercial available; however, simply adding a mixing system to a storage tank with a high hydraulic retention time will generally not improve water quality. Monitoring and modeling studies can aid in understanding storage cycling and mixing.

Routine Inspections

It is important that utilities have formal procedures for routine inspections of storage tanks. Periodic (for example, weekly) ground-based inspections may include inspection of sanitary, safety, and security conditions. Regular (such as quarterly) inspections may examine areas that are not visible from the ground (roof, interior), as described in Kirmeyer (1999). A comprehensive storage facility inspection should be scheduled annually to assess the condition of tank and coatings, corrosion control, and any safety, sanitary, or structural deficiencies. It is important that utility storage facility inspection schedules and procedures comply with any local regulatory requirements.

The AWWA Manual M42—Steel Water Storage Tanks (AWWA 2013b) outlines procedures for both internal and external tank inspections. If the storage facility is drained for inspection, ANSI/AWWA Standard C652 (ANSI/AWWA C652-11) provides guidelines for disinfection before placing back in service. ANSI/AWWA C652 also provides guidelines for equipment disinfection if divers or remote operational vehicles are used during the inspection process. If the tank is painted or a coating applied, the facility should not be placed back in service until tested for any volatile organic compounds.

Security and Telecommunications Equipment

Storage facilities are critical infrastructures for maintaining distribution system security, as well as water quality. The Water Infrastructure Security Enhancements (WISE) program provides guidance for protecting these facilities and other water assets. Steps to improve security can include considering the installation of:

- Locked hatches
- Fencing around the tank site
- Locked gates
- Elevated tank ladders
- Vandal deterrent shields or safety cages at the bottom of the ladders
- Modifications to vents, overflows, and other interior tank penetrations to minimize access to the tank interior
- Video surveillance systems

Where tanks have installed telecommunication equipment, the utility should have written guidelines for installation and methods of attachment, to ensure that they will not affect the structural integrity of the tank. Maintenance of the telecommunication equipment must not compromise the security of the tank.

Status

The status of storage facility operation and maintenance practices, as related to achieving a desired level of distribution system performance, may be assessed by reviewing the following self-assessment questions. The self-assessment team is not limited to these questions and may consider discussing additional topics

Table 3-29. References for storage facility operation and maintenance

Reference	Description
AWWA, 2014. ANSI/AWWA D102-14, Coating Steel Water-Storage Tanks. Denver, Colo.: AWWA.	Describes coating systems for inside and outside steel tanks.
AWWA, 2009. ANSI/AWWA D103-09, Factory-Coated Bolted Carbon Steel Tanks for Water Storage. Denver, Colo.: AWWA.	Provides minimum requirements for the design, construction, inspection, and testing of new factory coated, bolted carbon steel tanks.
AWWA, 2011. ANSI/AWWA D104-11, Automatically Controlled, Impressed-Current Cathodic Protection for the Interior Submerged Surfaces of Steel Water Storage Tanks. Denver, Colo.: AWWA.	Describes automatically controlled, impressed-current cathodic protection for the interior of steel water tanks.
AWWA, 2011. ANSI/AWWA C652-11 Disinfection of Water Storage Facilities. Denver, Colo.: AWWA.	Addresses disinfection of tanks and disinfection of divers or remote operational vehicles.
AWWA/EES (Economic and Engineering Services), 2002. *Finished Water Storage Facilities.*	Provides a review of state requirements for storage tank inspection.
AWWA, 2013b. M42, *Steel Water-Storage Tanks, Revised Edition.* Denver, Colo.: AWWA.	Outlines procedures for internal and external tank inspections.
Friedman, M.; Kirmeyer, G.; Pierson, G.; Harrison, S.; Martel, K.; Sandvig, A.; & Hanson, A., 2005. AwwaRF #2875. Development of Distribution System Water Quality Optimization Plans. Denver, Colo.: AWWA Research Foundation.	Provides an example of a comprehensive management plan for finished water storage facilities.
Grayman, W.M.; Rossman, L.A.; Arnold, C.; Deininger, R.A.; Smith, C.; Smith, J.F.; & Schnipke, R., 2000. AwwaRF #260. Water Quality Modeling of Distribution System Storage Facilities. Denver, Colo.: AWWA Research Foundation.	Provides water utilities with tools to assist in the design, retrofit, and operation of distribution system storage facilities to minimize poor mixing and excessive water age.

Reference	Description
Kirmeyer, G.K.; Kirby, L.; Murphy, B.M.; Noran, P.F.; Martel, K.D.; Lund, T.W.; Anderson, J.L.; & Medhurst, R., 1999. AwwaRF #254. Maintaining Water Quality in Finished Water Storage Facilities. Denver, Colo.: AWWA Research Foundation.	Recommends a five-step approach to establishing standard operating procedures to address storage facility water quality.
Kirmeyer, G.J.; Friedman, M.; Clement, J.; Sandvig, A.; Noran, P.; Martel, K.; Smith, D.; LeChevallier, M.; Volk, C.; Antoun, E.; Hiltebrand, D.; Dyksen, J.; & Cushing, R., 2000. AwwaRF #90798. Guidance Manual for Maintaining Distribution System Water Quality. Denver, Colo.: AWWA Research Foundation.	Provides recommendations to minimize the detention time of water in storage facilities, to maintain positive pressure, and to bulk water direction and velocity control.
Mahmood, F.; Pimblett, J.; Hill, C.; & Chowdhury, Z., 2009. CFD Modeling and Spreadsheet Tool to Evaluate Mixing and Water Quality in Storage Tanks. Proc. AWWA Annual Conference, San Diego, Calif.	Developed a spreadsheet tool to conduct a general desktop mixing evaluation of storage tanks. The results were used to select tanks for additional computational fluid dynamics modeling and temperature profiling.
Roberts, P.J.W.; Tian, X.; Sotiropoulos, F.; & Duer, M., 2006. AwwaRF #2898. Physical Modeling of Mixing in Water Storage Tanks. Denver, Colo.: AWWA Research Foundation.	Scale model studies used three-dimensional laser-induced fluorescence (3DLIF) to demonstrate effects of various inlet and outlet configurations on storage mixing.

related to storage facility operation and maintenance that may assist in identifying and addressing performance-limiting factors.

- Are storage facilities routinely (at least weekly) sampled for disinfectant residual testing? Are special sampling surveys conducted periodically to assess residual uniformity?
  - ▲ Special sampling surveys may include surveys such as an analysis of water quality at varying depths through the storage tank.
  - ▲ To ensure that water quality in the storage tank is monitored, samples collected at the storage tank outlet are recommended to be collected during the tank's draw (discharge) cycle.
- Is water use from the storage facility monitored? Is water age in the storage facility tracked and controlled? Is the tank turnover rate optimized for water age consideration?
  - ▲ Hydraulic modeling and computational fluid dynamics studies can be useful tools for optimization of this area.
- Is a routine inspection of all storage facilities conducted annually or according to regulatory requirements (if more frequent)? Are formal procedures for routine inspections in place and required maintenance identified and addressed?

- Are all storage facilities cleaned at least every 5 years? Is service interrupted when facilities are isolated for cleaning?

Action

If areas of the utility's storage facility operation and maintenance practices are considered to have a status of Partially Optimized or Not Optimized, and this performance improvement variable has been prioritized and selected for improvement, the following steps are recommended to be considered in the development of an action plan to improve performance in this area.

- If storage facilities are not routinely sampled for disinfectant residual, consider developing a plan to include storage facilities in the utility's disinfectant residual monitoring plan. It is optimal to sample as many storage facilities as possible. At minimum, storage facilities at the highest risk for water quality degradation should be addressed.

  - Consider the benefits of developing special sampling surveys to evaluate water quality in storage tanks most as risk for water quality degradation. Special sampling surveys may include an evaluation of water quality at varying depths within the storage tank.

  - If water quality degradation is identified, implement steps to improve water quality. These may include actions ranging from installing a mixing system in the tank to making overall changes in distribution system operation to minimize water age and improve water quality.

- If water use from the storage facility is not monitored or water age in the storage facility is not tracked and controlled, consider incorporating these components into the distribution system hydraulic model. A hydraulic model can be beneficial for projecting water age throughout the system under different conditions and identifying areas that may be at risk for experiencing increased water age. Additional information about the use of hydraulic models is provided in chapter 4.

- If storage tanks are not inspected on a routine basis or if SOPs do not exist for periodic and/or regular inspection, develop a schedule and procedures for storage facility inspection. Ensure that the schedule and procedures developed for inspection comply with all local regulatory requirements. Inspection procedures are recommended to include follow-up steps to take if issues are identified during the inspection process that need to be address to maintain the integrity of the storage facility.

- If storage facilities are not regularly cleaned, develop a schedule and procedures for cleaning storage facilities. If this task is contracted to an outside entity, ensure that it is incorporated into the utility budget for future planning. If service is interrupted when facilities are isolated for cleaning, determine if alternatives may exist that allow for continued service throughout the cleaning process.

Performance-Limiting Factor Summary

Table 3-30 summarizes factors related to storage facility operation and maintenance practices that may be limiting optimized performance of the distribution system. Check whether these factors are Optimized and Documented, Partially Optimized, or Not Optimized. Factors identified as Partially Optimized or Not Optimized will be prioritized in chapter 7, Identification and Prioritization of Performance-Limiting Factors, where an action plan can be developed and implemented that allows for optimization of parameters selected for improvement.

Table 3-30. Storage facility operation and maintenance

Self-assessment category	Questions for gauging optimization status	Optimized and documented	Partially optimized	Not optimized	Comments
Disinfectant Residuals	Are storage facilities routinely (at least weekly) sampled for disinfectant residual testing? Are special sampling surveys conducted periodically to assess residual uniformity?				
Storage Facility Water Use	Is water use from the storage facility monitored? Is water age in the storage facility tracked and controlled? Is the tank turnover rate optimized for water age consideration?				
Inspection	Is a routine inspection of all storage facilities conducted annually or according to regulatory requirements (if more frequent)? Are formal procedures for routine inspections in place and required maintenance identified and addressed?				Routine inspections are described above and are mainly external.
Cleaning	Are all storage facilities cleaned at least every 5 years? Is service interrupted when facilities are isolated for cleaning?				Evaluate cleaning frequency depending on inspection results.

WATER AGE, MODELING

Understanding

Water age is a major factor influencing water quality deterioration within the distribution system. The two main mechanisms for water quality decline are interactions with the pipe wall and reactions within the bulk water itself. As the bulk water travels through the distribution system, it undergoes various chemical, physical, and aesthetic transformations that affect water quality. These transformations advance to a greater or lesser extent depending on flow and temperature (AWWA/EES 2002a).

Additional references related to water age and modeling can be found in Table 3-31. The self-assessment questions for this Performance Improvement Variable are included in Table 3-32.

Components of an Optimized System for Water Age Management

Water age cannot be easily monitored, and there is no "ideal" water age that is suitable for all systems. Water age can be estimated using hydraulic models, tracer studies, computational fluid dynamic modeling, and spatial sampling (Walski 2003, Brandt 2004, Mahmood 2009). Water quality models can be used with hydraulic models to predict concentrations of chlorine, DBPs, and other constituents in a distribution system (Rossman et al. 1994, Vasconcelos et al. 1996).

For each nonconservative water quality parmeters, rates and mechanisms of reaction (decay and growth) must be identified and measured for a particular system. Water quality parameters, which change at least partially because of water age, can be directly measured. These include parameters such as chlorine residual, nitrification-related parameters, temperature, DBPs, and certain aesthetic characteristics (such as color, tastes, odors).

Systems with stable finished waters (characterized by low nutrient levels, low temperature, noncorrosive, and with minimal accumulations of biofilm or sediments) will likely tolerate a higher water age before water quality degradation is likely to occur. The extent of water age impacts on water quality will be affected by the disinfectant type and dose, system piping and storage facilities operation and maintenance (Kirmeyer et al. 2000, AWWA/EES 2002a, GLUMRB 2003, Walski et al. 2003, Brandt et al. 2004, Mahmood 2009, AWWA G200-15).

Effective water age management (and modeling) involves the following:

- Committing qualified staff and resources to review distribution system and storage configurations. Objectives should include the verification of design features that favor optimal water turnover rates (such as cycling, mixing, flow), while still addressing fire flow capacity and system pressure requirements. Properly calibrated hydraulic models can be helpful for understanding the impacts of system design on water age and for evaluating the effectiveness of operational or design modifications at reducing water age. Specific modeling and evaluation techniques may be used to better understand flow patterns within storage facilities and their impacts on water age.

- Establishing surveillance testing schedules at key locations in the system for water age and water quality indicator parameters (such as chlorine residual, pH, temperature, HPCs, DBPs, or others, depending on system and source water characteristics). Key locations include downstream of storage facilities (when water is being withdrawn), areas in which large volume piping serves low demand consumers, and system extremities, including dead-ends and pressure zone boundaries. Sampling downstream of storage facilities should be conducted when the facility is draining and should represent the range of water age conditions in storage. Special studies, such as continuous monitoring, depth or spatial sampling, or computational fluid dynamics or scale modeling, may be needed to profile water age within storage facilities.

- Implementing maintenance practices (such as implementing unidirectional flushing programs or rehabilitating or replacing deteriorating pipelines) that can reduce the system's vulnerability to water age-related water quality impacts.

Table 3-31. References for effective water age management and modeling

Reference	Description
AWWA/EES (Economic and Engineering Services), 2002. *Effects of Water Age on Distribution System Water Quality*.	Reference includes good summary descriptions of factors and tools to manage water age; cites examples of water age and indicator factors.
AWWA, 2015. G200-15, Distribution Systems Operation and Management. Denver, Colo.: AWWA.	Systems should provide good storage turnover/establish target turnover rate. Storage tank monitoring is recommended.
Brandt, M.; Clement, J.; & Powell, J., 2004. AwwaRF #91006F. Managing Distribution Retention Time to Improve Water Quality, Phase I & Phase II. Denver, Colo.: AWWA Research Foundation.	Guide development involved 23 utilities and includes 68 problem solution cases. WQPs of most concern: 1) discoloration, 2) disinfectant residual (DBP or nitrification related), 3) microbial, 4) taste and odor. Most water age solutions need common sense reasoning. Phase II includes an electronic guidance manual to determine water age and to solve water quality problems related to water age.

Reference	Description
Great Lakes Upper Mississippi River Board of State Public Health and Environmental Managers (GLUMRB), 2003. *Recommended Standards for Water Works (aka 10 States Standards).*	*Storage 7.0.2 Location of reservoirs.* Consideration should be given to maintaining water quality when locating water storage facilities. *7.0.6 Stored Water Turnover.* The system should be designed to facilitate turnover of water in the reservoir. Consideration should be given to separate inlet and outlet pipes, baffle walls, or other acceptable means to avoid stagnation.
Kirmeyer, G.J.; Friedman, M.; Clement, J.; Sandvig, A.; Noran, P.; Martel, K.; Smith, D.; LeChevallier, M.; Volk, C.; Antoun, E.; Hiltebrand, D.; Dyksen, J.; & Cushing, R., 2000. AwwaRF #90798. Guidance Manual for Maintaining Distribution System Water Quality. Denver, Colo.: AWWA Research Foundation.	Provides operational guidelines to minimize the water's detention time, to maintain positive pressure, and to control the bulk water direction and velocity. Describes potential causes of WQ degradation with recommended preventative O&M and design to prevent or remediate. Field validation studies conducted in 19 systems. Cites various recommended turnover rates, for example, daily storage turnover of 20-50% or full turnover every 3-5 days.
Mahmood, F.; Pimblett, J.; Hill, C.; & Chowdhury, Z., 2009. CFD Modeling and Spreadsheet Tool to Evaluate Mixing and Water Quality in Storage Tanks. Proc. AWWA Annual Conference, San Diego, Calif.	Developed a spreadsheet tool to conduct a general desktop mixing and cycling evaluation of storage tanks.
USEPA, 2006. Distribution System Indicators of Water Quality. Total Coliform Rule Issue Paper.	Temperature difference between storage tanks and entry to the distribution system can suggest stratification in storage tanks and therefore degradation of water quality that could lead to microbial regrowth in the distribution system.
Walski, T.M.; Chase, D.V.; Savic, D.A.; Grayman, W.; Beckwith, S.; Koelle, E., 2003. *Advanced Water Distribution Modeling and Management.* Waterbury, Ct.: Haestad Press.	Textbook provides extensive explanation of drinking water distribution system modeling concepts and examples.

- Training utility personnel to recognize and respond to customer complaints or routine system water quality monitoring results, which may suggest that water age-related water quality degradation has occurred.

Status

The status of water age management and modeling as related to achieving a desired level of distribution system performance may be assessed by reviewing the following self-assessment questions. The self-assessment team is not limited to these questions and may consider discussing additional topics related to water age management and modeling that may assist in identifying and addressing performance-limiting factors.

- Does the system have known areas where water age is high? Are these areas closely monitored for disinfectant residual, microbial parameters, and DBPs?
- Is water age monitored using a calibrated hydraulic model and current operating conditions? Is the system operated to minimize water age?
- Does the system have a calibrated hydraulic model that includes water quality parameters? Is the model used regularly and recalibrated as changes are made?

Action

If areas of the utility's water age management and modelling practices are considered to have a status of Partially Optimized or Not Optimized, and this performance improvement variable has been prioritized and selected for improvement, the following steps are recommended to be considered in the development of an action plan to improve performance in this area.

- If areas of known high water age are not regularly monitored for water quality parameters, such as disinfectant residual, microbial parameters, and DBPs, consider modifying the utility's distribution system water quality sampling plan to incorporate targeted sampling of high water-age areas.
  - ▲ If high water-age areas have not yet been identified in the distribution system (according to the utility's definition of high water age), utility staff should consider identifying those locations most at risk for experiencing high water age and incorporating these sites into the distribution system sampling plan.
- If a hydraulic model is not currently used to monitor water age, consider the benefits of developing a hydraulic model that may be used for this purpose and to enable the system to operate in a manner that helps to minimize water age.
- If the system does not have a hydraulic model that incorporates water quality parameters, consider the benefits of developing such a model.
  - ▲ Whether or not the hydraulic model contains water quality parameters , it is essential that the hydraulic is regularly recalibrated to help ensure accuracy. It is also important that access to the hydraulic model (or to an individual who can operate the hydraulic model) is provided to staff that require it. Additional information about hydraulic modeling is included in chapter 4.

Performance-Limiting Factor Summary

Table 3-32 summarizes factors related to water age management and modelling practices that may be limiting optimized performance of the distribution system. Check whether these factors are Optimized and Documented, Partially Optimized, or Not Optimized. Factors identified as Partially Optimized or Not Optimized will be prioritized in chapter 7, Identification and Prioritization of Performance-Limiting Factors, where an action plan can be developed and implemented that allows for optimization of parameters selected for improvement.

Table 3-32. Water age management and modeling

Self-assessment category	Questions for gauging optimization status	Optimization status			
		Optimized and documented	Partially optimized	Not optimized	Comments
Water Age Tracking	Does the system have known areas where water age is high? Are these areas closely monitored for disinfectant residual, microbial parameters, and DBPs?				
Water Age Control	Is water age monitored using a calibrated hydraulic model and current operating conditions? Is the system operated to minimize water age?				Assess the use of the hydraulic model to identify excessive water age areas and situation.
Hydraulic Model	Does the system have a calibrated hydraulic model that includes water quality parameters? Is the model used regularly and recalibrated as changes are made?				

WATER LOSS CONTROL

Understanding

Utilities benefit from quantifying and addressing two types of water losses: apparent and real. Apparent water losses are especially costly to water utilities because they result in lost revenues, either from unauthorized consumption (such as theft) or inaccurate metering or data handling. Real losses result from leakage from mains, services, or tank overflows and result in higher operating costs for water production and pumping. Additionally, if the utility's sources of supply are strained, real losses can cause the utility to invest substantial sums for additional supply, treatment, and pumping facilities. An effective water loss control program involves a multidisciplinary effort to account for all water that enters a distribution system, evaluate the types of losses, and implement activities to reduce the losses. Distribution systems optimized with respect to water loss control complete a water audit on a regular basis to characterize and quantify the water loss that may be occurring in the distribution system. The completion of a water audit may be required by regulation in some areas. Utilities should comply with all applicable regulations with respect to water audits and water loss control.

Additional references relating to water loss control can be found in Table 3-33. The self-assessment questions for this Performance Improvement Variable are included in Table 3-34.

Components of an Optimized System for Water Loss Control

Metering of Production Flows and Customer Consumption

Effectively accounting for water loss in a distribution system starts with accurate, reliable flow metering of all sources of water entering the distribution system and exiting at points of consumption. Ideally, flow metering should also be provided at all transfer points between sub-systems, pressure zones, or locations where there is the ability to segment the distribution system into district metered areas (DMAs). DMAs can be created within the water system by providing meters at pressure-reducing and pumping stations. Other likely metering locations to create DMAs are where there are only one or two transmission mains crossing a natural boundary (such as a river) or an artificial boundary (such as an interstate highway). Fanner et al. (2007) provides additional guidance for establishing DMAs and monitoring zone flows.

Universal customer metering is also essential for accounting for water losses. Proper meter selection, installation, maintenance, and a defined meter replacement schedule are essential to ensuring accurate and reliable flow data. Maintenance of meters should include calibration and verification, performed according to the manufacturer's recommendations. Guidance can be obtained from meter manufacturers as well as in AWWA Manuals of Practice M6—Water Meters-Selection, Installation, Testing, and Maintenance (AWWA 2012a) and M22—Sizing Water Service Lines and Meter (AWWA 2014a). Meters removed from use should be periodically tested to ensure the utility's meter replacement schedule is effective. Utilities must also ensure that they are in compliance with local requirements for the testing and reporting of meters removed from use.

Annual Water Audit

Long-term accountability for water is established by conducting an annual water audit. AWWA Manual of Practice M36—Water Audits and Loss Control

Programs (AWWA 2016b) provides detailed instructions, templates, and guidance for completing the audit. To download free AWWA water audit software, visit www.awwa.org, which provides a Microsoft Excel–based program that provides a tool for utilities to complete a water audit. This program helps to classify the various types of consumption or losses. Trending and analyzing the annual results over a multiyear horizon is important, so that water loss control activities can be optimized. As water audits are becoming increasingly mandatory in many localities, the utility must stay aware of new and changing regulations that will affect their water audits. For example, many utilities are now required to perform annual water audits as part of conservation efforts. Some utilities are even required to have their annual water audits validated by a third party to ensure the accuracy of the audit.

Real-Time Hydraulic Data Collection and Monitoring

The utility should have a monitoring system in place that allows for timely recognition of changes in distribution system flows and pressures. Continuous monitoring is most effectively provided by a supervisory control and data acquisition system that is configured with alarm set points and the ability to automatically trend flow rates and pressures. Conditions at tanks, pump stations, and other key points in the distribution system should be monitored and compared with expected values to identify abnormalities that may be indicative of a loss of system integrity, such as a water loss event. When a water loss event is suspected, the utility needs to investigate the causes to find the source and apply the appropriate response to minimize water loss, as well as the potential for contaminant intrusion.

Measures to Control Water Losses

The utility should set targets and develop strategies to control water losses, both real and apparent. In addition to the flow metering and testing described previously, the other important elements to control apparent losses include:

- Analysis of customer consumption history to identify any aberration from normal baseline usage patterns
- Field visits to customer premises to investigate abnormal billing records
- Quality control procedures for verifying proper data handling of meter reading and billing information
- Enforcement actions to stop and deter unauthorized consumption.

Four major factors that may be considered to address control of real losses include: (1) active leakage control, (2) timeliness and quality of repairs, (3) pressure management, and (4) distribution system asset management. An important decision to select the most appropriate strategies is determining the utility's economic level of leakage (ELL). The ELL recognizes that some degree of water loss

Table 3-33. References for development of water loss control programs

Reference	Description
Andrews, L.; Gasner, K.; Sturm, R.; Jernigan, W.; Cavanaugh, S.; & Kunkel, G., 2017. WRF#4639b. Utility Water Audit Validation: Principles and Programs. Denver, Colo.: Water Research Foundation.	Research report summarizing the fundamentals of water audit validation to provide readers with a technical understanding of water audit data validation and its applications.
AWWA, 2015. G200-15, Distribution Systems Operation and Management. Denver, Colo.: AWWA.	Establishes a standard with minimum requirements for tracking water loss and leakage.
AWWA, 2005. *Water Distribution System Assessment Workbook*. Smith, C. editor. Denver, Colo.: AWWA.	Provides checklists for assessing application and maintenance of meters and leak detection in water distribution systems.
AWWA, 2006n. M33, *Flowmeters in Water Supply*, 2nd ed. Denver, Colo.: AWWA.	Provides background information, procedures, forms, and diagrams for the proper selection, installation, inspection, testing, and maintenance of meters used in water supply facilities.
AWWA, 2012a. *M6—Water Meters-Selection, Installation, Testing, and Maintenance*, 5th ed. Denver, Colo.: AWWA.	Provides background information, procedures, forms, and diagrams for the proper selection, installation, inspection, testing, and repair of customer meters.
AWWA, 2016b. M36, *Water Audits and Loss Control Programs, Fourth Edition*. Denver, Colo.: AWWA.	Provides background information, guidelines, procedures, forms, and diagrams for establishing a water loss control program, conducting a water audit, and understanding and controlling real and apparent losses.
Fanner, M.; Thornton, J.; Liemberger, R.; & Sturm, R., 2007. AwwaRF #91163. Evaluating Water Loss and Planning Loss Reduction Strategies. Denver, Colo.: AWWA Research Foundation.	Provides extensive information about defining, quantifying, and controlling water loss based on application of methods used in the United Kingdom to North American utilities.
Fanner, M.; Thornton, J.; Liemberger, R.; Sturm, R.; Davis, S.; & Hoogerwerf, T., 2007. AwwaRF #91180. Leakage Management Technologies. Denver, Colo.: AWWA Research Foundation.	Provides extensive information about leak detection equipment and techniques, analytical methods and tools, and water loss control strategy development based on international best practices and pilot programs in North American utilities.
Sturm, R.; Gasner, K.; Wilson, T.; Preston, S.; & Dickinson, M.A., 2014. WRF #4372a Real Loss Component Analysis: A Tool for Economic Water Loss Control. Denver, Colo.: Water Research Foundation.	Provides an analysis of the AWWA Water Audit Software that is available to utilities to use for leakage analysis.

is unavoidable, and it may be financially impractical to reduce real losses below a certain level. Two reports by Fanner (2007) and the AWWA Manual of Practice M36 (AWWA 2016b) describe various methods to control water loss and provide guidance with regard to determining the ELL.

Status

The status of utility water loss control practices as related to achieving a desired level of distribution system performance may be assessed by reviewing the following self-assessment questions. The self-assessment team is not limited to these questions and may consider discussing additional topics related to water loss control that may assist in identifying and addressing performance-limiting factors.

- Does the system use the AWWA/International Water Association (IWA) water audit method described in AWWA Manual M36? Does the system calculate and track the Infrastructure Condition Factor, the Infrastructure Leakage Index (ILI), and the annual real losses? Is a system water audit performed annually? If so, is the water audit third-party validated?

- Is the system divided into DMAs or pressure zones to optimize leak detection? Does the system have an active leak detection program (acoustic or other)?

- Does the utility have a defined meter replacement schedule? Is meter accuracy verified before new meters are installed? Are meters removed from use tested to verify that the meter replacement schedule is effective? If advanced metering infrastructure (AMI) is in place, is it used to notify consumers of possible private leaks?

Action

If areas of the utility's water loss control practices are considered to have a status of Partially Optimized or Not Optimized, and this performance improvement variable has been prioritized and selected for improvement, the following steps are recommended to be considered in the development of an action plan to improve performance in this area.

- If the system does not use the AWWA/IWA water audit method described in AWWA Manual M36—or does not perform a water audit, consider developing procedures that enable the utility to complete an annual water audit, incorporating these procedures. As previously described, AWWA provides software resources for water auditing.

  ▲ Consider the benefits of calculating and tracking the Infrastructure Condition Factor, ILI, and annual real losses in conjunction with this.

  ▲ Utilities may also consider the benefits of having a water audit validated by a third party.

  ▲ Utilities should comply with all applicable regulations regarding water auditing.

- If the system does not have an active leak detection program, consider taking steps to initiate such a program, concentrating on the areas at highest potential risk for leaks.
  - ▲ If the system is not currently divided into DMAs, consider whether DMAs would be beneficial in providing additional information that may assist the utility's optimization efforts.
- If the system does not have a defined meter replacement schedule, develop a schedule for meter replacement that incorporates accuracy verification.
  - ▲ Consider evaluating and using AMI data to proactively information customers about the potential presence of leaks.

Performance-Limiting Factor Summary

Table 3-34 summarizes factors related to water loss control practices that may be limiting optimized performance of the distribution system. Check whether these factors are Optimized and Documented, Partially Optimized, or Not Optimized. Factors identified as Partially Optimized or Not Optimized will be prioritized

Table 3-34. Water loss control

Self-assessment category	Questions for gauging optimization status	Optimized and documented	Partially optimized	Not optimized	Comments
Water Audit	Does the system use the AWWA/International Water Association (IWA) water audit method described in AWWA Manual M36? Does the system calculate and track the Infrastructure Condition Factor, the Infrastructure Leakage Index (ILI), and the annual real losses? Is a system water audit performed annually? If so, is the water audit third-party validated?				Calculate the ICF, ILI, and real loss volume and compare to past results. Review water audit procedures to ensure they conform to AWWA M36.
Leak Identification	Is the system divided into DMAs or pressure zones to optimize leak detection? Does the system have an active leak detection program (acoustic or other)?				

Self-assessment category	Questions for gauging optimization status	Optimization status			
		Optimized and documented	Partially optimized	Not optimized	Comments
Metering Accuracy	Does the utility have a defined meter replacement schedule? Is meter accuracy verified before new meters are installed? Are meters removed from use tested to verify that the meter replacement schedule is effective? If AMI is in place, is it used to notify consumers of possible private leaks?				

in chapter 7, Identification and Prioritization of Performance-Limiting Factors, where an action plan can be developed and implemented that allows for optimization of parameters selected for improvement.

WATER QUALITY SAMPLING AND RESPONSE
Understanding

Water quality sampling is essential for the operation of all distribution drinking water systems. The best strategy for water quality sampling is to have an optimized sampling plan and to operate in accordance with the plan. This sampling plan should enable representative sampling of the water in the distribution system and should include both compliance and nonregulatory monitoring. Maintenance of water quality integrity in the distribution system will require operators (field, treatment, water quality analyst) to be diligent in monitoring. Several federal and local drinking water quality regulations require monitoring for disinfectant residual and other water quality parameters within the distribution system, according to approved monitoring plans. The following information is provided to help utilities assess their own water quality sampling programs. General tips and references for conducting effective monitoring are summarized in several documents (CFR Title 40; Kirmeyer et al. 2000, 2002; USEPA 2006f, 2008a). Measures described here may be conducted by utilities or contract laboratory staff. If contracted, utilities must provide oversight for contractor activities. More detailed information about water quality sampling is provided in chapter 5.

Additional references related to water quality sampling and response can be found in Table 3-35. The self-assessment questions for this variable are included in Table 3-36.

Components of an Optimized System for Water Quality Sampling and Response

Distribution system sampling is required for regulatory compliance, including the monitoring requirements associated with the following USEPA regulations:

- Revised Total Coliform Rule (RTCR)
- Disinfectants and Disinfection By-products Rule (D/DBPR)
- Surface Water Treatment Rule (SWTR),
- Groundwater Rule (GWR)
- Lead and Copper Rule (LCR)

Monitoring requirements (locations, frequencies, analytical methods, reporting) for compliance sampling are specified in the respective rules (CFR Title 40). Outside of the United States, many distribution system regulations also typically incorporate specific monitoring requirements into their components. Although the contents of this guide focus primarily on USEPA regulations, utilities are encouraged to contact their local regulatory agency for information about the monitoring requirements associated with specific regulations.

An effective sampling plan should establish a sufficient number of fixed or dedicated locations for compliance purposes or for performing special evaluations. The plan should describe sample collection, preservation, and storage (if required) methods appropriate to the parameter to be tested. Data should preferably be incorporated into electronic formats to facilitate trending. Sampling plans should include evaluation of data collected by both grab sampling and on-line monitors.

Selection of routine and special-purpose sampling sites should consider the following factors:

- Geographic dispersion
- Representation of various water sources
- Required sampling frequency
- Site accessibility (daily, weekly)
- System size
- Coverage of all pressure zones
- Storage tanks
- Critical users (such as hospitals, schools, storage facilities)
- Seasonal use
- Sampling in proportion to system demands and populations served

Suggested optimized sampling locations for disinfectant residual samples are provided in chapter 2. Known problem areas should also be monitored routinely with established trigger levels and actions for utility response. Routine samples from storage facilities should be representative of water leaving the storage facility under normal operating cycles and should represent normal stored water levels. Chemical treatment points in the distribution system should also be regularly monitored for process control purposes. Monitoring at nonroutine sampling locations includes areas upstream and downstream of main breaks and repairs, sampling associated with other system events or changes (such as maintenance, cleaning), and at customer complaint locations.

Except when sampling under the LCR requirements, sample taps (dedicated, hose-bibs, home taps) should be flushed before sample is taken. This is to ensure that the water from the main, at the service address, is represented in the sample. Dedicated sampling stations can be useful when allowed by the local regulatory agency. Sites, taps, and equipment used for collection of microbial samples should be kept free of external or internal contamination (no aerators or strainers, no hoses, no leaking packing material, and use of sterile sample collection technique).

Operators and analysts assigned to sample collection should be trained in all applicable sampling procedures and field methods. It can also be beneficial to verify the effectiveness of training, through the use of visual observation or standards analysis. Each water utility should have a written protocol for training, which addresses the following topics:

- Initial training and refresher training topics
- Tap flushing/sanitization—varies depending on parameter to be sampled
- Field parameter measurement
- Sample collection (bottle type, preservatives, dechlorinating agent, fill volume)
- Sample storage and transport
- Chain of custody and documentation
- Notification plan

Initial training (with demonstrations or video and review of written sampling procedures) should be provided for samplers and analysts. The technician should understand the purpose of water sampling and analysis and factors affecting sample or test result integrity. Approved, certified samplers are required in some areas for compliance sampling, and certified analysts and methods are usually required for several parameters. Certifications are not required for surveillance-type sampling; however, the sampler should be knowledgeable of proper sample handling techniques and water quality integrity triggers. Partnership for Safe Water subscribers submit a disinfectant residual sampling and analysis SOP as a component of the self-assessment completion report that is submitted to the program for peer review.

Results review and follow-up communications and investigations should be timely and appropriate. Review of multiple parameters or trend analysis may be needed to reveal patterns in distribution system water quality that may be indicative of events requiring response. Irregular sample results should be a basis for immediate response. In the case of irregular results, consider repeating the analysis to determine whether the irregularity may have been based on analytical error. Compliance sample result irregularities must be addressed according to regulatory requirements. Results of all sampling should be communicated to consumers or regulatory agency according to regulatory requirements. A water quality emergency notification plan should be established to provide a rapid and effective means of informing the public of contaminated, substandard, or unsafe conditions of the water. Water quality data trends can also be useful to identify areas or system components in need of maintenance or replacement (Kirmeyer et al. 2000).

Status

The status of water quality sampling and response procedures as related to achieving a desired level of distribution system performance may be assessed by reviewing the following self-assessment questions. The self-assessment team is not limited to these questions and may consider discussing additional topics related to water quality sampling and response that may assist in identifying and addressing performance-limiting factors.

- Is there a routine sampling and testing plan that includes monitoring beyond regulatory requirements? Does the plan address system-specific issues?

- Are routine samples (at least weekly) taken from all storage tanks (and facilities) and tested for microbial parameters, disinfectant residual, and other appropriate parameters?

- Is an SOP available and followed for main break or leak repair, including any required sampling and testing?

Action

If areas of the utility's water quality sampling and response procedures are considered to have a status of Partially Optimized or Not Optimized, and this performance improvement variable has been prioritized and selected for improvement,

Table 3-35. References for effective water quality sampling and response

Reference	Description
AWWA, 2012c. *Standard Methods for Examination of Water and Wastewater.* 22nd ed. E.W. Rice, R.B. Baird, A.D. Eaton, and L.S. Clesceri, editors. ISBN 978-0-87553-013-0.	Extensive listing of USEPA-approved analytical methods with detailed descriptions, including interferences.
CFR Title 40—Chapter 1, Subchapter. D.	Web site contains links for federal drinking water regulations, which describe types of samples required, locations, frequencies, analytical methods, and reporting for compliance sampling.
Kirmeyer, G.J.; Friedman, M.; Martel, K.; Thompson, T.; Sandvig, A.; Clement, J.; & Frey, M., 2002. AwwaRF #90882. Guidance Manual for Monitoring Distribution System Water Quality. Denver, Colo.: AWWA Research Foundation.	Matrix for prioritizing distribution system monitoring objectives. Matrix contains parameters grouped into: raw water; point of entry; representative distribution sites, and regulatory sites: pH, alkalinity, coliform, HPC, DBPs, nitrate, etc. are listed by monitoring objectives such as: nitrification; customer complaints.
Kirmeyer, G.J.; Friedman, M.; Clement, J.; Sandvig, A.; Noran, P.; Martel, K.; Smith, D.; LeChevallier, M.; Volk, C.; Antoun, E.; Hiltebrand, D.; Dyksen, J.; & Cushing. R., 2000. AwwaRF #90798. Guidance Manual for Maintaining Distribution System Water Quality. Denver, Colo.: AWWA Research Foundation.	Describes data uses to prioritize rehabilitation and replacement of cast iron, flushing needed, differentiate between source water and distribution system tastes and odors, monitor complaints by service area and water source, track complaint trends.
USEPA, 2008a. Total Coliform Rule/Distribution System Federal Advisory Committee (TCRDS FAC). Agreement in Principle.	Document describes changes recommended for the revised total coliform rule including tiered assessments to be conducted in response to coliform occurrences.
USEPA, 2006f. Interactive Sampling Guide for Drinking Water System Operators EPA816-F-03-016.	Interactive training DVD explains sampling requirements and methods for test required under USEPA regulations; uses short videos and PowerPoint presentations.
USEPA, 2008f. Water Quality in Small Community Distribution Systems Reference Guide for Operators EPA/600/R-08/039CD	Guidance manual for treatment and distribution operations; developed for small systems.

the following steps are recommended to be considered in the development of an action plan to improve performance in this area.

- If a routine sampling and testing plan does not exist, take steps to create distribution system sampling plan. The plan should incorporate the collection of samples for both regulatory and nonregulatory requirements, including samples that may address any system-specific needs. It is recommended that a sampling plan includes the following:

- ▲ Sampling sites
- ▲ Sample collection frequency
- ▲ Specific required sample collection conditions or procedures
- ▲ Parameters to be tested
- ▲ Identification of who is performing the sampling
- ▲ Additional information about the development of a sampling plan is included in chapter 5.

- If routine (weekly) samples are not collected from storage tanks and tested for disinfectant residual, microbial parameters, and other appropriate parameters, consider incorporating these locations into the system's sampling plan. At a minimum, it is recommended to include the sites that may exhibit the greatest risk for water quality degradation.

- If an SOP is not available for main break/leak repair, including required sampling and testing, develop such an SOP so that procedures are followed consistently by utility staff.

- ▲ Consider training staff on any new procedures to ensure learning, and develop a method to evaluate the effectiveness of the training.

Performance-Limiting Factor Summary

Table 3-36 summarizes factors related to water quality sampling and response that may be limiting optimized performance of the distribution system. Check whether these factors are Optimized and Documented, Partially Optimized, or Not Optimized. Factors identified as Partially Optimized or Not Optimized will be prioritized in chapter 7, Identification and Prioritization of Performance-Limiting Factors, where an action plan can be developed and implemented that allows for optimization of parameters selected for improvement.

Table 3-36. Water quality sampling and response

Self-assessment category	Questions for gauging optimization status	Optimization status			
		Optimized and documented	Partially optimized	Not optimized	Comments
Water Quality Sampling and Testing	Is there a routine sampling and testing plan that includes monitoring beyond regulatory requirements? Does the plan address system-specific issues?				
Storage Tank Testing	Are routine samples (at least weekly) taken from all storage tanks (and facilities) and tested for microbial parameters, disinfectant residual, and other appropriate parameters?				
Main Break or Leak Repair Testing	Is an SOP available and followed for main break or leak repair, including any required sampling and testing?				Review the main break testing procedures to ensure they are followed.

CHAPTER 4

DESIGN EVALUATION

Introduction

The focus of this chapter is to assess design-related performance-limiting factors. The data gathered from system records and the completed system asset inventory provide the basic information needed to assess design-related performance-limiting factors. This chapter focuses on system design and optimization using distribution system modelling as well as on the characteristics of physical facilities and equipment including size, materials of construction, age, type, flexibility, and redundancy.

The self-assessment of physical facilities and equipment concentrates on the materials of construction and their attributes (age, size) that may affect water quality and service reliability. Other design related factors affecting water quality are covered in the performance improvement variables chapter (chapter 3). This chapter focuses on the evaluation of system design in the following areas: distribution system design tools (computer models) and asset management for physical assets including pipeline materials, storage facilities materials and construction, pumping facilities, and valves and hydrants. *Although age is not necessarily the definitive indicator for infrastructure condition, this attribute is frequently used to narrow and focus the assessment.* Partnership for Safe Water (PSW) subscriber utilities should note that this chapter contains several information summary tables, the completion of which is a required component of the self-assessment process. PSW subscribers receive blank copies of these tables in the Self-Assessment Worksheets document, received upon joining the program. A table of relevant references pertaining to distribution system asset management and design (Table 4-11) is provided at the end of this chapter.

DISTRIBUTION SYSTEM DESIGN TOOLS (COMPUTER MODELS)

Understanding

There are two main types of distribution system computer models used by many utilities: (1) a hydraulic model and (2) a water quality model. The hydraulic model is built on and contains features such as pressures and flows at maximum and average demand conditions. Traditionally, these models reflect a snapshot of the operation; however, advances in modeling technology have led to operation that is also able to reflect an extended period of operation (extended period simulation [EPS]). The water quality model is built using a hydraulic model EPS base, incorporating elements to enable predictions of water quality changes. Typical water quality parameters included in these models include source water tracing, water age, and disinfectant residual.

Although distribution system modelling can have certain limitations, it is a valuable tool for planning, design, and operation. Examples of how distribution system modelling may be applied for planning, design, and operational purposes are provided next. Distribution system models can provide water utilities the needed understanding of system design and operational requirements for service reliability. Distribution system models that include both hydraulic and water quality features can enhance utility optimization efforts.

Planning

- Long-range capital improvement plans benefit from the availability of hydraulic models. These are used to identify system expansion for population growth and to replace aging infrastructure.

- Hydraulic models are helpful when evaluating the effects of conservation measures on projected demand.

- The effects of future main rehabilitation are evaluated using hydraulic models.

- Distribution reservoirs (storage) are best located and sized by first evaluating various alternatives using a hydraulic model, especially a calibrated model.

- Initial system development plans are reviewed using the water quality model. Often, this results in modifications that can improve finished water storage design.

Engineering Design

- Use simulations of fire flows from various hydrant locations to demonstrate compliance with fire protection standards. Deficiencies can be identified and corrected using hydraulic models with EPS.

- Flow studies can help evaluate situations to select the proper main size for a location.
- Hydraulic models are useful for evaluating flows to a reservoir to determine an optimum size for a location.
- Pump characteristics can be used to match pressure and head requirements with system demands.
- Pressures and flows at locations of interest are calculated by hydraulic models. This enables system operators to locate low pressure areas and to take action to address the issue.
- Hydraulic models are used to locate pressure zone boundaries to optimize flow delivery and improve energy efficiency.

System Operation

- Hydraulic models (all hydraulic models in this section include EPS) can be used to train system operators by allowing them to experiment with different demand simulations.
- System problems can be diagnosed using the hydraulic model.
- Hydraulic models can be used to estimate the water loss from main breaks or quantifying smaller leaks for water system audits.
- Emergency operation scenarios can be simulated, using the hydraulic model, and responses practiced by system operators.
- The effect of load shifting between multiple treatment plants can be evaluated.
- A calibrated hydraulic model can be used by operating personnel to identify the cause of past operational problems.
- Hydraulic models can be used to help design an efficient flushing program.
- Hydraulic models are used to determine the best method to isolate a tank or system area to conduct maintenance activities.
- Water quality models can track possible contaminants introduced into a system, allowing operators to make adjustments to isolate the affected water and minimize the impact of the contamination.
- Water quality models are used to help operators to minimize water age and water quality deterioration by managing water age in the distribution system.
- Water quality models are used to determine the best sample locations for targeted water quality parameters.
- Water quality models can be used for source tracing. This is useful when investigating water quality concerns.

Features of an Optimized Distribution System Model

A fully optimized distribution system model is calibrated using field data from many points within the distribution system. The model includes water quality data for a variety of parameters of interest including, at a minimum, chlorine or chloramine residual decay and water age. The model (some may be based on the latest version of EPANET) may also incorporate some of all of the following features:

- Steady-state analyses
- Extended-period simulations
- Graphical user interface
- Error reporting
- Selective reporting
- System component characteristics (such as pump curves and flow- and pressure-regulating devices)
- Data management
- Automated fire flow calculations
- Scenario generation
- Water quality modeling and predictive capability

Once constructed and calibrated, it is important that the water utility establish a process that ensures that the distribution system model is routinely updated. A system model, similar to the water distribution system, is not a static entity. If water mains, pumps, or storage facilities are replaced, expanded, or otherwise modified, steps must to taken to adjust the hydraulic model to reflect current field conditions. A system model that relies on information stored in a geographic information system (GIS) can automatically reflect the current distribution system attributes that are maintained in the GIS. Hydraulic models are often complex. In many cases, the development and maintenance of models is the responsibility of the utility's engineering staff or an outside consultant retained for this objective. However, as described previously, models can be very beneficial to distribution system operation and are a tool that should be used by utility staff, as appropriate.

Status

The status of the utility's distribution system design tools, as they relate to achieving a desired level of distribution system performance, may be assessed by reviewing the following self-assessment questions. The self-assessment team is not limited to these questions and may consider discussing additional topics related to distribution system design tools that may assist in identifying and addressing performance-limiting factors.

- Does the utility have a distribution system hydraulic model?
- Does the hydraulic model incorporate the following characteristics?
  - ▲ Water age information
  - ▲ Disinfectant residual (chlorine or chloramine) decay information
- Is the output from the model compared with real system data for validation purposes?
- Does a process exist to periodically update the hydraulic model to reflect the dynamic nature of the distribution system?

Action

If areas of the utility's distribution system design tools are considered to have a status of Partially Optimized or Not Optimized, and this parameter has been prioritized and selected for improvement, the following steps are recommended to be considered in the development of an action plan to improve performance in this area.

- If a distribution system hydraulic model does not exist, consider the benefits of developing a distribution system model that can be used to support distribution system operational practices.
- In the development of a new model or the refinement of an existing model, consider incorporating, at minimum, information pertaining to distribution system water age and disinfectant residual decay. These parameters can be beneficial in predicting the potential for water quality degradation in high-risk areas of the system.
- For a hydraulic model to be as applicable as possible, it is important that its degree of accuracy is known. If the accuracy of the model has not been verified by the utility, consider completing the following steps:
  - ▲ Compare the output from the model against real system data to validate the performance of the model.
  - ▲ Consider calibrating the hydraulic model on a periodic basis, so that the model represents the true nature of the system as accurately as possible.

Performance-Limiting Factors Summary

Table 4-1 summarizes factors related to distribution system design tools that may be limiting optimized performance of the distribution system. Check whether these factors are Optimized and Documented, Partially Optimized, or Not Optimized. Factors identified as Partially Optimized or Not Optimized will be prioritized in chapter 7, Identification and Prioritization of Performance-Limiting Factors, where an action plan can be developed and implemented that allows for optimization of parameters selected for improvement.

ASSET MANAGEMENT

Understanding

Utilities with optimized distribution systems use asset management to evaluate and record the physical characteristics of distribution system assets, such as pipelines, pumping facilities, storage tanks, valves, hydrants, and other system components. An asset management program is a beneficial tool applied by utilities to help determine the optimum replacement and rehabilitation frequency for distribution system assets. For utilities with established asset management programs, much of the information addressed in this section, and in the self-assessment questions that follow, will be readily available. For utilities without active asset management programs, this section will serve as a good starting point for establishing a program, which may be an action item outcome of the self-assessment process.

Table 4-1.   Distribution system design tools assessment

Self-assessment category	Questions for gauging optimization status	Optimized and documented	Partially optimized	Not optimized	Comments
Distribution System Design Tools	Does the utility have a distribution system hydraulic model?				
	Does the hydraulic model incorporate the following characteristics: • Water age information • Disinfectant residual (chlorine or chloramine) decay information				
	Is the output from the model compared with real system data for validation purposes?				
	Does a process exist to periodically update the hydraulic model to reflect the dynamic nature of the distribution system?				

Components of an Optimized Asset Management Program

Optimized asset management programs include many of the following common elements (Grigg 2004):

- Distribution system inventory
- Condition assessment
- Maintenance program
- System planning and needs assessment
- System financial accounting
- Asset management framework
- Capital improvement plan and program
- Capital budget
- Capital management system

The US Environmental Protection Agency (USEPA) has developed a Check-Up Program for Small Systems (CUPSS, https://www.epa.gov/dwcapacity/information-check-program-small-systems-cupss-asset-management-tool). This software program has an asset inventory as its basis. Small systems can use this program to create an asset inventory and to assess the priority for repairs and replacement of individual components. The program contains useful life estimates for each asset type that are based on experience and Government Accounting Standards Board financial tables. The results from this program can also be used to complete this self-assessment.

The USEPA has also created a publication titled "Asset Management: A Best Practices Guide" (USEPA 2008g). This guide includes the description of best practices for determining the current state of system assets, including practices related to the following topics:

- Preparing an asset inventory and system map
- Developing a condition assessment and rating system
- Assessing remaining useful life by consulting projected useful life tables or decay curves
- Determining asset values and replacement costs

The core of all asset management programs is the distribution system asset inventory and condition assessment. Based largely on these results, priorities are established for repair and replacement of major system components. GIS is an excellent tool to map the water distribution system infrastructure and track operations and maintenance histories on distribution system assets. GIS can also be effectively linked to distribution system models and computerized maintenance management systems.

Several factors influence utility rehabilitation and replacement decision-making including age, condition, materials, cost, level of service, asset redundancy, safety, and criticality. A comprehensive asset management program contains information related to all of these factors. For the purposes of this self-assessment, age and materials are used to highlight components for further evaluation.

Status

The status of the utility's asset management practices, as they relate to achieving a desired level of distribution system performance, may be assessed by reviewing the following self-assessment questions. The self-assessment team is not limited to these questions and may consider discussing additional topics related to asset management that may assist in identifying and addressing performance-limiting factors.

- Does the distribution system have an up-to-date schematic map that indicates all major physical assets including distribution mains, storage facilities, pumps, hydrants, valves, and meters?
    - Maps should indicate the normal position of valves (open or closed).
    - Capacity of storage facilities should be shown.
    - Pumping station capacities should be noted.
- Does a complete inventory of all distribution system assets exist? Information that is recommended to be recorded in an asset inventory includes:
    - Age of asset
    - Materials of construction for each asset (pipes, storage tanks, pumps, valves, hydrants, etc.)
    - Length and size of asset (for example, pipe diameter and length). Note whether these measurements are estimated or actual.
- Are maintenance records kept that include maintenance procedures performed, maintenance dates, and asset condition observations made at the time maintenance is performed?
    - Maintenance procedures and dates
    - Observations at time of maintenance: asset condition
- Does an asset management program exist that incorporates the following elements?
    - Condition assessment
    - Residual life determination

- ▲ Replacement cost estimate
- ▲ Established level of service targets
- ▲ Criticality determination
- ▲ Development of a capital improvement plan

Action

If areas of the utility's asset management program are considered to have a status of Partially Optimized or Not Optimized, and this parameter has been prioritized and selected for improvement, the following steps are recommended to be considered in the development of an action plan to improve performance in this area.

- If an asset inventory does not exist or does not contain sufficient detail to optimally manage the utility's physical assets, consider developing an asset inventory or adding the ability to capture additional details that may assist in the asset management process.
  - ▲ Ensure that the asset inventory is updated on a regular basis so that it accurately reflects the changing nature of the distribution system.
- If an accurate distribution system schematic does not exist, create a schematic that includes location and details about distribution system physical aspects. Consider including water quality sampling sites in the distribution system schematic.
- If an asset management program does not exist, consider developing a comprehensive asset management program that incorporates the elements described previously in this section.
  - ▲ Small systems may consider using the USEPA CUPSS resources as tools to support development of an asset management program.
  - ▲ Consider including maintenance record information in the asset management program or related database.

Performance-Limiting Factors Summary

Table 4-2 summarizes factors related to asset management that may be limiting optimized performance of the distribution system. Check whether these factors are Optimized and Documented, Partially Optimized, or Not Optimized. Factors identified as Partially Optimized or Not Optimized will be prioritized in chapter 7, Identification and Prioritization of Performance-Limiting Factors, where an action plan can be developed and implemented that allows for optimization of parameters selected for improvement.

PIPELINE MATERIALS

Although pipelines are included as a component of a comprehensive asset management plan, a separate discussion and self-assessment questions are provided because of the importance of this particular asset category.

Understanding

Distribution main pipelines are a major portion of a utility's infrastructure. Drinking water is conveyed to users through a system of pipelines, with the goal of consistently delivering high-quality water to customers. The ages of pipes, materials of construction, and pipeline condition are some of the primary factors affecting the ability of pipelines to achieve this goal.

The oldest cast iron pipes from the late 19th century are typically described as having an average useful lifespan of about 120 years because of their pipe wall

Table 4-2. Asset management assessment

Self-assessment category	Questions for gauging optimization status	Optimization status			
		Optimized and documented	Partially optimized	Not optimized	Comments
Asset Management	Does the distribution system have an up-to-date schematic map that indicates all major physical assets including distribution mains, storage facilities, pumps, hydrants, valves, and meters?				
	Does a complete inventory of all distribution system assets exist?				
	Are maintenance records kept that include maintenance procedures performed, maintenance dates, and asset condition observations made at the time maintenance is performed?				
	Does an asset management program exist that incorporates the following elements? • Condition assessment • Residual life determination • Replacement cost estimate • Established level of service targets • Criticality determination • Development of a capital improvement plan				

thickness (NASTB 2006). In the 1920s, the manufacture of iron pipes changed to improve pipe strength, but the changes also produced a thinner wall. These pipes have an average life of about 100 years. Pipe manufacturing continued to evolve in the 1950s and 1960s with the introduction of ductile iron pipe that is stronger than cast iron and more resistant to corrosion.

Polyvinyl chloride (PVC) pipes were introduced in the 1970s and high-density polyethylene in the 1990s. Both of these are very resistant to corrosion, but they do not have the strength of ductile iron. Post-World War II pipes typically have an average life of 75 years (NASTB 2006). A variety of pipeline materials exist, including asbestos cement (AC), reinforced concrete (RC), prestressed concrete cylinder (PCCP), steel, and (HDPE) high-density polyethylene. A distribution system may contain a variety of pipeline materials, a factor that should be taken into consideration when evaluating the development of pipeline rehabilitation or replacement plant.

When considering rehabilitation or replacement of a pipeline several factors are usually considered, including:

- Age
- Material of construction
- Main break data
- Soil conditions
- Water stability
- Cost

Main break data that includes observations of pipe condition are often the determining factor in pipeline rehabilitation or replacement decisions; however, older pipes are often correlated with increased main break frequency. Additionally, certain pipe materials are more susceptible to corrosion and, thus, exhibit higher break frequency.

Optimized distribution systems aim to minimize the amount of unlined metal pipelines, because these are often the source of water quality problems and have proven to be susceptible to increased failure rates. Pipeline rehabilitation and replacement programs are often based on failure frequency and water quality concerns. Failure events and frequencies should, therefore, be tracked and pipelines exhibiting the highest failure rates should be programmed for replacement. An optimized system schedules pipeline replacements to reduce the failure rate and enhance water quality. Systems should also not overlook opportunities to work synergistically with their communities, with regard to the replacement of high-risk pipeline at times when roadways may be scheduled for repair, facilitating access to the pipelines that lay beneath.

Status

The status of the utility's pipeline materials, as they relate to achieving a desired level of distribution system performance, may be assessed by reviewing the following self-assessment questions. The self-assessment team is not limited to these questions and may consider discussing additional topics related to pipeline materials that may assist in identifying and addressing performance-limiting factors.

- Is there a complete distribution system pipeline inventory that includes information about pipeline size, length, age, location, and materials of construction?
- Does the utility's asset management plan include specific criteria that are used to determine pipeline replacement?
- Has the utility determined the total length and materials of construction for distribution system pipeline that is more than 75 years old?
- Has the utility identified all areas in which unlined metal pipe is present in the system?

Although it is a beneficial exercise for all readers, PSW subscribers complete and submit Table 4-3 as a component of the distribution system self-assessment completion report. An electronic version of Table 4-3 is provided to all Partnership subscribers upon joining the program, in the Self-Assessment Worksheets document. Table 4-3 is designed to capture relevant information about distribution system pipe that is older than 75 years; however, it may also be used to collect beneficial information about pipeline of any age.

Action

If areas of the utility's pipeline materials assessment are considered to have a status of Partially Optimized or Not Optimized, and this parameter has been prioritized and selected for improvement, the following steps are recommended to be considered in the development of an action plan to improve performance in this area. It is recommended that utilities develop a pipeline rehabilitation and

Table 4-3.   Pipe inventory summary (> 75 years)

Pipe material	Length (miles)	Estimated	Documented
Unlined cast or ductile iron			
Lined cast or ductile iron			
Concrete (all types)			
Polyvinyl chloride (PVC)			
Polyethylene (PE)			
Other			

replacement plan to reduce the unlined metal pipe inventory over time and replace older pipes according to a prioritized strategy incorporated into the asset management program.

- If a complete distribution system pipeline inventory does not exist, consider developing one or adding it to an existing asset management program. The inventory should include pertinent information about pipeline size, materials of construction, age, location, condition, and other pertinent factors that may influence utility pipeline rehabilitation or replacement decisions.

  ▲ As a component of a pipeline inventory, utility staff should have knowledge of areas in which unlined metal pipe is installed, as well as areas of the system containing pipes that are more than 75 years of age.

- If criteria have not been established that drive pipeline rehabilitation and/or replacement activities, consider developing such criteria, considering that pipeline age is not the sole factor driving pipeline rehabilitation and replacement efforts.

Performance-Limiting Factor Summary

Table 4-4 summarizes factors related to pipeline materials that may be limiting optimized performance of the distribution system. Check whether these factors are Optimized and Documented, Partially Optimized, or Not Optimized. Factors identified as Partially Optimized or Not Optimized will be prioritized in chapter 7, Identification and Prioritization of Performance-Limiting Factors, where an action plan can be developed and implemented that allows for optimization of parameters selected for improvement.

STORAGE FACILITES: MATERIALS AND CONSTRUCTION

Although storage facilities are included as a component of a comprehensive asset management plan, a separate discussion is provided due to the importance of this particular asset category.

Understanding

Distribution system storage facilities may take many forms. They can be constructed at ground level, elevated, or underground, and can be fabricated from metal, plastic, concrete, or masonry. Some are coated internally and some are protected externally from corrosion. Induced current corrosion protection is even

Table 4-4.   Pipeline materials assessment

Self-assessment category	Questions for gauging optimization status	Optimization status			
		Optimized and documented	Partially optimized	Not optimized	Comments
Pipeline Materials Assessment	Is there a complete distribution system pipeline inventory that includes information about pipeline size, length, age, location, and materials of construction?				Describe criteria used to prioritize replacement and repair.
	Does the utility's asset management plan include specific criteria that are used to determine pipeline replacement?				
	Has the utility determined the total length and materials of construction for distribution system pipeline that is more than 75 years old?				
	Has the utility identified all areas in which unlined metal pipe is present in the system?				

provided for some metallic installations. Sizes can vary from a few hundred gallons to millions of gallons.

System storage is necessary to mitigate fluctuations in water demand. Demand changes are caused by seasonal influences, population density, large users, industrial activity, and fire flow. Most water distribution systems have large storage requirements due to these factors.

Storage tanks often become potential sources of water quality degradation. A number of factors may contribute to this consequence. Water age, lack of mixing, postprecipitation and sediment accumulation, and internal and external corrosion are among the main influences on water quality in storage facilities. The age of the storage tank does not always mean that water quality will suffer; however, surface reactions often contribute to corrosion and other factors influencing water quality in older tanks. It is therefore important to identify older storage facilities and assess the need for replacement or rehabilitation. The USEPA CUPSS asset management program lists the useful life of storage tanks as 30 years, so tanks approaching this age and beyond may be considered in this evaluation. However, many tanks can and do exhibit a functional life longer than this, so age should not be the sole factor driving storage facility replacement and/or rehabilitation efforts.

Optimized distribution systems have a defined inspection and maintenance program for all storage tanks. Storage tank maintenance for optimized distribution systems was discussed in greater detail in chapter 3. Note that regulatory requirements exist in many areas with respect to storage tank inspection, which should supersede any recommendations provided in this guidance. As part of their asset management program, optimized systems determine and prioritize tank replacement according to a variety of factors.

Status

The status of the utility's storage facilities, as they relate to achieving a desired level of distribution system performance, may be assessed by reviewing the following self-assessment questions. The self-assessment team is not limited to these questions and may consider discussing additional topics related to storage facilities that may assist in identifying and addressing performance-limiting factors.

- Has the size, age, and materials of construction been identified for all distribution system storage tanks?
- Has a condition assessment been performed for all storage facilities?
- If the system contains tanks that are greater than 30 years old, has a rehabilitation or replacement plan been developed to address these tanks?
    - ▲ If it was determined that a rehabilitation or replacement plan was not necessary, provide the rationale to support this determination.

Although it is a beneficial exercise for all readers, PSW subscribers complete and submit Table 4-5 as a component of the distribution system self-assessment completion report. An electronic version of Table 4-5 is provided to all Partnership subscribers, upon joining the program. Table 4-5 is designed to capture relevant information about the system's storage facility inventory.

Action

If areas of the utility's storage facilities assessment are considered to have a status of Partially Optimized or Not Optimized, and this parameter has been prioritized and selected for improvement, the following steps are recommended to be considered in the development of an action plan to improve performance in this area. It is recommended that utilities develop a storage facility rehabilitation and replacement plan to reduce the number of older facilities in the system, over time, and replace older tanks according to a prioritized strategy incorporated into the asset management program.

- If the characteristics of distribution system storage facilities have not been defined, develop a plan to collect and document key storage facility

Table 4-5.   Storage facility inventory

Storage tank ID	Type (elevated, ground, or underground)	Storage capacity (MG—note if alternate units are reported)	Material of construction	Age	Corrosion protection (active, passive, or none)	Internal and external inspection date and results

characteristics. Such information may be recorded in the utility's asset management database.

- If a condition assessment has not been performed to evaluate storage tank condition, develop a plan to complete a condition assessment. In developing such a plan, consider prioritizing the tanks that may be at the greatest risk for deterioration.

- If rehabilitation and/or replacement plans have not been considered for the system's oldest storage tanks (for example, those greater than 30 years of age), consider developing a formal rehabilitation or replacement strategy.

  - ▲ Age is not the sole factor in determining storage facility rehabilitation and/or replacement needs. If older tanks are determined not to require rehabilitation or replacement at the current time, consider identifying drivers that may influence the need for future replacement and a regular inspection schedule so that future needs may be proactively identified.

Performance-Limiting Factors Summary

Table 4-6 summarizes factors related to storage facilities that may be limiting optimized performance of the distribution system. Check whether these factors are Optimized and Documented, Partially Optimized, or Not Optimized. Factors identified as Partially Optimized or Not Optimized will be prioritized in chapter 7, Identification and Prioritization of Performance-limiting Factors, where an action plan can be developed and implemented that allows for optimization of parameters selected for improvement.

PUMPING FACILITIES

Although pumping facilities are included in a comprehensive asset management plan, a separate discussion of this topic is provided due to the importance of this particular asset category.

Understanding

Distribution system pumping facilities provide needed pressure and flow throughout the distribution network. Centrifugal pumps are the most common pumps, but many other types of pumps are also used for this purpose, depending on the specific nature of the distribution system. Most distribution system pumps are steel construction, but some are lined to resist corrosion. Other materials are often used for some of the pump components. Some systems also use hydropneumatic tanks to maintain pressure.

In some cases, pumps are critical in maintaining the hydraulic integrity of the system. Positive pressure is required (> 20 psi) to prevent contaminant intrusion and ensure water quality. Redundant pumping capacity is necessary to maintain system pressure if there is a pump malfunction. Many systems are equipped with multiple pumps of various capacities to provide the needed backup security for emergency situations such as fires or major line breaks.

The age of the pump does not always mean that it is in imminent danger of failure; however, older pumps often require more maintenance and may be off-line more frequently. It is therefore important to identify older pumping facilities (> 20 years) and to assess the need for replacement. Older pumps may also

Table 4-6. Storage facility assessment

Self-assessment category	Questions for gauging optimization status	Optimization status			
		Optimized and documented	Partially optimized	Not optimized	Comments
Storage Facility Assessment	Has the size, age, and materials of construction been identified for all distribution system storage tanks?				Report tanks that are currently scheduled for replacement or repair.
	Has a condition assessment been performed for all storage facilities?				
	If the system contains tanks that are greater than 30 years old, has a rehabilitation or replacement plan been developed to address these tanks?				

operate less efficiently than newer pumps. The topic of pump efficiency testing and how it relates to energy management is included in chapter 3.

Optimized systems should have a defined operation and maintenance program for their pumping facilities. Data should be tracked on the percentage of time spent on predictive and preventive maintenance compared with reactive (corrective) repairs due to pump system breakdowns. This ratio is referred to as the Planned Maintenance Ratio. The objective for most utilities is to maximize the Planned Maintenance Ratio, such that the majority of maintenance performed on equipment is planned, rather than corrective.

The AWWA benchmarking performance indicator survey (AWWA 2016e) contains reported results for the Planned Maintenance, which is calculated as described by the equation below:

Planned Maintenance Ratio (%) = 100 [(planned maintenance hours)/(planned and corrective maintenance hours)]

AWWA's 2013 benchmarking survey indicates that the 75th percentile value for the Planned Maintenance Ratios reported by water utilities and combined systems were 77% and 75%, respectively. Well-maintained systems may exceed 80% for the planned maintenance ratio reported for pump maintenance.

As part of the asset management program, water utilities determine and prioritize pump replacement according to a variety of factors, including asset age. Distribution system pump planning also considers the need for excess pumping capacity to ensure the minimum pressure goals of the system (> 20 psi) are met continuously under worst case conditions (the largest pumping unit out of service on a maximum demand day).

Status

The status of the utility's pumping facilities, as they relate to achieving a desired level of distribution system performance, may be assessed by reviewing the following self-assessment questions. The self-assessment team is not limited to these questions and may consider discussing additional topics related to pumping facilities that may assist in identifying and addressing performance-limiting factors.

- Does the utility have an inventory of distribution system pumping facilities, including information related to capacity, number and size of pumps, age of pumps/installation date, inspection results, and strategy for redundancy?

- Has the utility developed standard operating procedures (SOPs) for pumping facility operation and maintenance that include predictive and preventive maintenance elements?

- Is the Planned Maintenance Ratio calculated and tracked for pumping facilities?

- Does the asset management plan include specific criteria for replacement and rehabilitation of pumps in its pump replacement/rehabilitation strategy?
- Does the system have adequate backup pumping capacity to have the firm capacity to meet the maximum date demand and a continuous minimum pressure > 20 psi?

Although it is a beneficial exercise for all readers, PSW subscribers complete and submit Table 4-7 as a component of the distribution system self-assessment completion report. An electronic version of Table 4-7 is provided to all Partnership subscribers upon joining the program. Table 4-7 is designed to capture relevant information about the system's pumping facility inventory.

Action

If areas of the utility's pumping facilities assessment are considered to have a status of Partially Optimized or Not Optimized, and this parameter has been prioritized and selected for improvement, the following steps are recommended to be considered in the development of an action plan to improve performance in this area. It is recommended that utilities develop a pumping facility rehabilitation and replacement plan to reduce the number of older facilities in the system, over time, and replace older pumps according to a prioritized strategy incorporated into the asset management program.

- If the utility does not have an inventory of all distribution system pumping facilities, develop an inventory, including the asset information previously referenced.
- If the utility does not have SOPs for pump operation and maintenance items, including preventive and predictive maintenance, develop SOPs that incorporate these elements. Ensure that all appropriate staff is trained in essential pump operation and maintenance procedures.
  - ▲ If SOPs for pump operation and maintenance exist, develop a schedule that helps to ensure periodic review and updating of SOPs, as appropriate.

Table 4-7. Pumping facility inventory summary

Name of pumping facility	Number of pumps and sizes	Designed capacity	Firm capacity (largest unit out of service)	Age of the oldest unit in facility

Table 4-8.   Pumping facility assessment

Self-assessment category	Questions for gauging optimization status	Optimization status			Comments
		Optimized and documented	Partially optimized	Not optimized	
Pumping Facilities	Does the utility have an inventory of distribution system pumping facilities, including information related to capacity, number and size of pumps, age of pumps/installation date, inspection results, and strategy for redundancy?				Report pumps that are currently scheduled for replacement or repair.
	Has the utility developed standard operating procedures (SOPs) for pumping facility operation and maintenance that include predictive and preventive maintenance elements?				
	Is the Planned Maintenance Ratio calculated and tracked for pumping facilities?				Report the Planned Maintenance Ratio for the 12-month reporting period. Describe the plan to improve this performance measure.
	Does the asset management plan include specific criteria for replacement and rehabilitation of pumps in its pump replacement/rehabilitation strategy?				
	Does the system have adequate backup pumping capacity to have the firm capacity to meet the maximum date demand and a continuous minimum pressure > 20 psi?				

- If the Planned Maintenance Ratio is not considered in pump evaluation, calculate the Planned Maintenance Ratio for at least the highest priority pumps to establish a performance baseline and quantify progress in this area.

- If sufficient pumping redundancy does not exist to allow the system to meet minimum pressure requirements under worst-case conditions, develop and implement a plan to provide the required level of redundancy.
  - ▲ Ensure that regulatory requirements for pump redundancy are met, if any apply.
  - ▲ Sufficient redundancy is considered to be a pump or combination of pumps that will meet peak demand with the largest unit out of service.
  - ▲ Also consider securing a redundant, backup power source to maintaining pumping capabilities in an emergency.
- If the asset management program does not include criteria for pump replacement—or the utility has not developed a comprehensive strategy for pump rehabilitation/replacement, take steps to develop a strategy and ensure that the asset management program has the ability to record the appropriate information related to pump replacement criteria.

Performance-Limiting Factors Summary

Table 4-8 summarizes factors related to pumping facilities that may be limiting optimized performance of the distribution system. Check whether these factors are Optimized and Documented, Partially Optimized, or Not Optimized. Factors identified as Partially Optimized or Not Optimized will be prioritized in chapter 7, Identification and Prioritization of Performance-Limiting Factors, where an action plan can be developed and implemented that allows for optimization of parameters selected for improvement.

VALVES AND HYDRANTS

Although valves and hydrants are included in a comprehensive asset management plan, a separate discussion of valves and hydrants is provided due to the importance of this particular asset category.

Understanding

Distribution system valves and hydrants are critical for efficient and safe system operation. Valves are necessary to isolate sections of the system, to direct flow, and sometimes to control pressure. Hydrants are provided for use in firefighting. Additionally, they are used to flush mains when needed, to fill mobile water tanks, and for other uses.

Valves are used in emergencies to redirect flow and to isolate distribution system areas when there are main breaks. It is crucial that these valves can be readily located and easily operated during a crisis. A valve exercise program is one way to help ensure that valves are able to be located and operable at the critical times when their proper operation is needed most. Although valves are available in a variety of materials, most are constructed of metal. Valve age does not necessarily determine the rate of failure or malfunction; however, age is one important factor to consider when conducting an evaluation for replacement.

Hydrants are most often used for firefighting. Because this is an emergency situation, it is critical that this asset functions reliably. Nearly all hydrants are of metal construction and therefore prone to corrosion. Hydrant age is again only one factor to consider when evaluating replacement.

The age of valves and hydrants does not always mean that they are in imminent danger of failure or malfunction; however, older valves often require more maintenance and may be less reliable. It is therefore important to identify older valves and hydrants and to determine if they are in need of replacement.

Optimized distribution systems should have a defined valve exercising and rehabilitation or replacement program and a hydrant inspection, replacement, and maintenance program (valve and hydrant maintenance and inspection are discussed in chapter 3). As part of the asset management program, valve and hydrant replacement may be prioritized according to a variety of factors.

Status

The status of the utility's valves and hydrants, as they relate to achieving a desired level of distribution system performance, may be assessed by reviewing the following self-assessment questions. The self-assessment team is not limited to these questions and may consider discussing additional topics related to valves and hydrants that may assist in identifying and addressing performance-limiting factors.

- Does the utility have a current (updated) valve and hydrant inventory? Do valve and hydrant records include installation date and inspection results?

- Does the utility's asset management plan include specific criteria for replacement and rehabilitation of valves and hydrants?

Although it is a beneficial exercise for all readers, PSW subscribers complete and submit Table 4-9 as a component of the distribution system self-assessment completion report. An electronic version of Table 4-9 is provided to all Partnership subscribers, upon joining the program. Table 4-9 is designed to capture relevant information about the system's valves and hydrants.

Action

If areas of the utility's valves and hydrants assessment are considered to have a status of Partially Optimized or Not Optimized, and this parameter has been

Table 4-9. Valve and hydrant inventory summary

Valves and hydrants	Total number	% > 35 years	% > 70 years
Valves < 10 in.			
Valves ≥ 10 in.			
Hydrants, wet barrel			
Hydrants, dry barrel			
Other valves			

prioritized and selected for improvement, the following steps are recommended to be considered in the development of an action plan to improve performance in this area. It is recommended that utilities develop a valve and hydrant rehabilitation and replacement plan to reduce the number of older facilities in the system, over time, and replace older valves and hydrants according to a prioritized strategy incorporated into the asset management program.

- If the utility does not have an inventory of valves and hydrants, does not have an updated inventory, or the inventory does not contain relevant information, such as installation date, age, inspection results, or valve direction, take steps to develop and/or update the inventory information.

- If the utility's asset management plan does not include specific criteria for valve rehabilitation/replacement, in addition to age, consider identifying factors that would drive the decision to rehabilitate or replace valves or hydrants, and ensure that this information is recorded and periodically reviewed.

Performance-Limiting Factors Summary

Table 4-10 summarizes factors related to valves and hydrants that may be limiting optimized performance of the distribution system. Check whether these factors are Optimized and Documented, Partially Optimized, or Not Optimized. Factors identified as Partially Optimized or Not Optimized will be prioritized in chapter 7, Identification and Prioritization of Performance-Limiting Factors, where an action plan can be developed and implemented that allows for optimization of parameters selected for improvement.

Table 4-10. Valves and hydrants assessment

Self-assessment category	Questions for gauging optimization status	Optimization status			Comments
		Optimized and documented	Partially optimized	Not optimized	
Valves and Hydrants	Does the utility have a current (up-dated) valve and hydrant inventory? Do valve and hydrant records include installation date and inspection results?				Report the date of the last inventory update.
	Does the utility's asset management plan include specific criteria for replacement and rehabilitation of valves and hydrants?				

Table 4-11. References for asset management and design

Reference	Description
AWWA, 2016e. *Benchmarking Performance Indicators for Water and Wastewater Utilities: 2016 Edition.* Denver, CO: AWWA. ISBN 9781625761972	AWWA's Benchmarking Surveys provide results for key utility performance indicators for water, wastewater, and combined utilities.
Grigg, N.S., 2004. AwwaRF #91025F. Assessment and Renewal of Water Distribution Systems. Denver, Colo.: AWWA Research Foundation.	Provides a description of the components of an optimized asset management system.
NASTB, 2006. Drinking Water Distribution Systems: Assessing and Reducing Risks. Washington, D.C.: National Research Council Report.	Provides information about long-term infrastructure viability in the United States.
USEPA, 2012. CUPSS: Check-Up Program for Small Systems.	Provides free asset management software and guidance, designed specifically for smaller utilities.
USEPA, 2008. Asset Management: A Best Practices Guide. USEPA 816-F-08-014.	Describes best practices for asset management and information about the benefits of developing and implementing a utility asset management program.

CHAPTER 5

APPLICATION OF OPERATIONAL CONCEPTS

Introduction

In the first parts of the self-assessment, system performance is evaluated and a determination is made of limitations that may be contributing to less-than-optimized performance. In this chapter, a critical operational factor is assessed: *application of operational concepts*. This chapter will focus specifically on operational control of all aspects of the distribution system because significant performance limitations can often exist in this area. The approach and operational methods used for maintaining system control can significantly affect distribution system performance, despite the existence of adequate infrastructure.

The assessment of the application of operational concepts focuses on sampling and associated system control testing, followed by data interpretation and associated operational adjustments. This chapter is divided into discussions of distribution system water quality control testing, recognizing performance deviations, system evaluation response, communication, and online instrumentation and supervisory control and data acquisition (SCADA).

The Partnership for Safe Water (PSW) goal for an optimized distribution system is to continually deliver high-quality water to all users. Operational control of the distribution system must be consistently and proactively applied to maintain the physical, hydraulic, and water quality integrity necessary to achieve this goal. System operators, managers, and engineers should work together to develop and implement a carefully designed monitoring plan to provide the data needed to detect and correct any threats to distribution system integrity. Timely application of operational concepts by knowledgeable and empowered operators is one characteristic of an optimized distribution system.

WATER QUALITY CONTROL TESTING

Understanding

For a system to accomplish the goal of consistently delivering high-quality water, it is necessary to optimize the performance of each optimization parameter and many of the performance improvement variables (chapters 2 and 3). This is important because a breakdown in performance anywhere in the system potentially increases the chance of viable pathogenic organisms or undesirable substances reaching the customers' taps.

To achieve the highest levels of performance for each optimization parameter, information focusing on water quality entering the system and on the performance of the distribution system must be *routinely* gathered and recorded. Based on this information, appropriate controls must then be exercised to maintain *consistent* water quality. The terms *routinely* and *consistent* are stressed because it is necessary to have adequate staff to monitor and control the operation. Information is continuously assessed so that process control adjustments can be made whenever water quality conditions dictate.

The foundation of an optimized distribution system is a comprehensive data set representative of that specific distribution system. This data set is then used by operators to make data-driven decisions affecting distribution system operations and performance. The first step in the development of a process control program is to establish commitment by staff at all levels to develop consistent performance goals for each key system component. The second, and equally important step, is to establish the tenacity to achieve these goals throughout the organization. For example, utility staff may choose to establish a goal of maintaining a 0.20 mg/L free chlorine residual at all sample locations throughout the distribution system and collect the data required to assess utility performance relative to that goal.

An example of typical process control targets for a midsized utility is shown in Table 5-1. Each PSW subscriber utility will create its own unique sampling guidelines and performance goals table for submission with the self-assessment completion report as part of the Self-Assessment Worksheet document.

After sampling is complete, operations staff should graph data on a trend chart to establish a baseline and visualize patterns in performance. Operations staff should review and interpret graphical data as often as possible, looking for unexpected deviations from baseline. Whenever appropriate, operators should discuss their observations, especially concerning water quality trends, with other operators and/or management. Operators should evaluate current water quality performance relative to the operational goals that have been established for the system. Part of this evaluation should include whether sufficient data exist to make an accurate determination of the system's current status. If data trends indicate that water quality goals may be compromised if the current operations are not altered, available system controls should be identified and implemented. A control is an adjustment that staff may implement that has the potential to

Table 5-1. Example operations guidelines outlining system performance goals

Subject:	Unit process performance goals		Guideline number:
Objective:	To establish process control targets for each sampling location or location type to ensure optimum system performance.		Date adopted: Date revised:
Sampling location or type	Tests		Target value
Entry Point(s) Water Quality	Turbidity pH Alkalinity Iron Manganese Odor Chlorine residual Chloride Sulfate TTHM HAA5 Total coliform		< 0.10 NTU 7.5-8.0 80-120 mg/L as $CaCO_3$ < 0.3 mg/L < 0.05 mg/L None 2.0-2.2 mg/L < 250 mg/L* < 250 mg/L* System specific[†] System specific[†] 0/100 mL
Distribution System Total Coliform Sites, DBP sites, and other routine (at least monthly scheduled) sample locations	Total coliform Chlorine residual TTHM HAA5 Turbidity		0/100 mL > 0.20 mg/L < 80 µg/L < 60 µg/L < 0.5 ntu
Storage Tanks & Reservoirs (Measured during both "fill" and "draw" cycles. Water quality goals should be met during draw cycle)	Turbidity pH Alkalinity Chlorine Water age		< 0.5 ntu < 7.6 80-100 mg/L as $CaCO_3$ > 0.5 mg/L System specific[‡]
Lead and Copper Rule Water Quality Parameter sites	pH Alkalinity Calcium Conductivity Chloride Sulfate		7-9 80-120 mg/L as $CaCO_3$ > 40 mg/L > 20 µS/cm § §
Continuous Monitoring Sites (for example, storage facilities and pump stations)	pH Total chlorine Free chlorine UV254 Conductivity Turbidity		Investigate changes from baseline. In the case of storage tank monitoring, comparison should be made between fill and draw cycles.
Water Quality Complaint Investigation Sites	Free chlorine Total coliform Temperature pH Odor‖ Color‖ Other‖		Representative of area 0/100 mL Representative of area

Notes:

Table is written for a free chlorine system. Systems using chloramine as a residual disinfectant should also monitor for nitrification-related parameters. Refer to chapter 3 for additional information regarding nitrification.

Definitions; $CaCO_3$, calcium carbonate; HAA5, sum of 5 haloacetic acids; TTHM, total trihalomethanes.

*Entry point chloride and sulfate concentrations listed are the USEPA Secondary Maximum Contaminant Levels. Systems may need to set utility-specific goals for corrosion control purposes. Refer to chapter 3 for additional information about internal corrosion control.

†Entry Point disinfection by-products (DBPs): Treatment staff should optimize in-plant treatment to the extent feasible and establish the lowest possible entry point DBP goal for the delivery of water to the distribution system. Water wholesalers may need to establish this goal in collaboration with their customers to help ensure compliance throughout the system.

‡Reduce water age as much as feasible while still maintaining adequate pressure and storage for fire protection.

§Operators should use their data to determine the optimal ratio of sulfate to chloride to enhance corrosion control. Refer to chapter 3 for additional information about internal corrosion control.

||Tests depending on the complaint, results should be representative of the area being tested.

impact system performance. Examples of controls that typically exist in most distribution systems include:

- Change the system pressure (reduce leakage, lower risk of breaks)
- Adjust the disinfectant residual
- Proactively flush areas with low water use before problems develop
- Manage water use from storage tanks to reduce water age and promote mixing
- Operate pumps to reduce pressure fluctuations
- Adjust flow patterns to balance velocity and reduce water quality impacts
- Control fire hydrant use to avoid pressure spikes
- Monitor and control entry point water quality to reduce the potential for degradation in the distribution system. An example of parameters that may be monitored for these purposes include pH, turbidity, corrosion inhibitor levels, chlorine to ammonia ratio for chloramine formation, iron and manganese, coagulant dosage optimization, disinfectant residual, and disinfection by-product (DBP) concentrations.
- Anticipate demand changes and communicate these to operators controlling entry-point supply. It is recommended to avoid making large changes in flow rate to meet system demands.

After available controls are identified, utility staff should reach a consensus decision regarding which control is most likely to remedy the problem. That control measure should be implemented, and appropriate follow-up water quality sampling should be conducted to document its actual impact. It is recommended that multiple control measures not be implemented simultaneously to address water quality issues because it can make it difficult to understand which control measure was most effective in addressing the issue. To most clearly understand the impact of a system adjustment, it is generally wise to make one adjustment at a time and conduct follow-up sampling after each adjustment has been made. At some point in the process, this may turn into a special study. System-specific special studies can provide extremely valuable data and information. Although special studies have significant value in optimizing distribution system performance, the Partnership also recognizes that quick decision-making by a properly certified operator can be necessary during emergency situations.

Sampling locations are selected based on developing a representative monitoring plan that adequately covers the various portions of the distribution system. In addition to considering regulatory requirements, the frequency of water quality sampling and testing is based on the importance of the measurement and the anticipated rate of change. Daily or weekly testing frequency is sufficient for many distribution system water quality tests. Some parameters may be tested monthly or quarterly (such as DBPs). Continuous monitoring is advantageous for monitoring parameters that may change on a frequent basis, sites that are not able to be regularly accessed for grab sampling, and to conduct special studies. Sampling locations should also include sites from all pressure zones as well as any historically problematic areas.

The established sampling frequency may not be sufficient, in all cases, to detect subtle performance problems (for example, the effects of biofilm growth). Insufficient data can cause operators to react to the problem, rather than implementing proactive corrective procedures. Sampling frequencies should be reviewed regularly to determine if they need to be changed. Be sure to review sampling procedures at any time that major changes are made to distribution system infrastructure or operation. The goal of sampling and testing is to detect and correct developing problems *before* they affect customers or system performance.

Examples of trend charts that may be used to evaluate distribution system performance include the system's daily minimum disinfectant residual, a frequency distribution of disinfectant residual concentration, and DBP levels at distribution system sampling sites. Stable performance would be indicated by level lines for these values. For data to be valuable, they need to be evaluated and interpreted on a regular basis and the necessary controls implemented to continuously maintain the desired performance goals. An example process control and sampling schedule for a distribution system is shown in Table 5-2. The utility will create its own unique process control sampling table for submission with the self-assessment completion report, as part of the Self-Assessment Worksheet document.

It is critical that administrative and operating personnel understand the importance of the system control program in attaining the utility's distribution system performance goals. Development of the performance goals by utility

Table 5-2. Example process control and sampling schedule for a distribution system (This table is intended to be used as an example. Distribution system personnel should customize this table based on the characteristics of the distribution system. Sampling frequencies also should be customized to capture seasonal differences or other key distribution system variables.)

Sample	Sample location	Tests	Frequency	Sampled by
System Entry Points	Treatment Plant Clearwell	Turbidity pH Alkalinity Flow TOC *Free chlorine—for free chlorine systems *Total chlorine *Monochloramine—for chloraminated systems *Free ammonia—for chloraminated systems TTHM HAA5 Temperature Total coliform	Continuous Continuous Daily Continuous Daily Continuous Continuous Monthly Monthly 1/Shift Daily	Treatment Plant Operator
Distribution System Total Coliform Sites	Designated System Locations	Temperature Chlorine residual Total coliform Heterotrophic plate count (HPC)	Weekly	Water Quality Technician
Distribution System Storage Tanks	Tank Outflow (when water is withdrawn)	Temperature Chlorine residual Total coliform	Weekly	Water Quality Technician
DBP Sites	Designated System Locations	Temperature Chlorine residual Total coliform TTHM HAA5 pH	Monthly (as required by regulations)	Water Quality Technician
Lead and Copper Rule Water Quality Parameter Sites	Designated System Locations	Temperature Chlorine residual Alkalinity pH Calcium Conductivity	Monthly (as required by regulations)	Water Quality Technician

Notes:

Disinfectant residual optimization goals and sampling sites are recommended to be established in accordance with the recommendations provided in chapter 2.

Definitions—HAA5, sum of 5 haloacetic acids; TTHM, total trihalomethanes, TOC, total organic carbon.

* When Table 5-2 refers to chlorine residual, the following parameters are recommended to be analyzed, and used for process control decision marking, depending on the nature of the primary disinfectant used in the distribution system:

• Free chlorine (free chlorine system)
• Total chlorine (free chlorine system, chloraminated system)
• Monochloramine (chloraminated system)
• Free ammonia (chloraminated system)

management, without input from the frontline operations staff, may not achieve the buy-in and acceptance necessary to attain these goals. Conversely, if frontline operators develop the program without management involvement, decisions may be made that are inconsistent with the utility mission. In any case, acceptance of system goals, by individuals throughout the organization, and a spirit of tenancy are required to successfully establish and apply system controls to improve performance.

All components of the system control program should be documented in Standard Operating Procedures (SOPs). These procedures should be written and tested by the operations staff. Once they are drafted, SOPs should be readily accessible to the entire operational staff. Accuracy of SOPs is critical, and SOPs need to be reviewed and revised on a regular basis. An annual review of SOPs is recommended to ensure the accuracy and relevancy of the documents. It is important to create an environment where SOPs are "living" documents and are meant to be updated and modified as new insights into the system control program are discovered.

Status

The status of water quality control testing as it relates to achieving a desired level of distribution system performance may be assessed by reviewing the following self-assessment questions. The self-assessment team is not limited to these questions and may consider discussing additional topics related to water quality control testing that may assist in identifying and addressing performance-limiting factors.

- Does the utility have a routine distribution system water quality testing program, including established water quality goals and sampling schedules?

- Are test results communicated to distribution system operators?

- Do operators make adjustments to the appropriate system controls based on the water quality test results?

- Has all of the operational staff been involved in the development of the water quality control testing program and SOPs, including emergency response guidelines?

Action

If areas of the utility's water quality control testing practices are considered to have a status of Partially Optimized or Not Optimized, and this parameter has been prioritized and selected for improvement, the following steps are recommended to be considered in the development of an action plan to improve performance in this area.

- If the utility does not have a routine distribution system water quality testing program, or has a program that does not allow for the collection of sufficient or representative data to assess distribution system performance, consider developing or enhancing the existing water quality testing program. Involve all appropriate utility staff in the process of developing the program, which will help to promote a culture of excellence and continuous improvement among system staff.

  ▲ High-priority parameters for consideration in the development of a distribution system water quality testing program include:)

    ■ Disinfectant residual (refer to chapter 2 for additional information about optimized sampling locations and performance goals)

    ■ Nitrification-related parameters (monochloramine, free ammonia) for chloraminated systems.

    ■ DBPs—total trihalomethanes (TTHM) and haloacetic acids (HAA5)

    ■ Flow

    ■ Lead and copper, ensuring (at minimum) compliance monitoring in accordance with current regulatory requirements. Note that, in reference to lead and copper, USEPA (as of the time of this writing) is currently developing long term revisions to the Lead and Copper Rule, which should be taken into account when finalized.

    ■ Corrosion-related parameters (refer to chapter 3 for an in-depth discussion of internal corrosion control and corrosion indicators)

    ■ Systems are encouraged to conduct additional special and investigative sampling for lead and copper, as well as other distribution system parameters, beyond minimum permit and regulatory requirements.

    ■ Systems should consider including point of entry water quality sampling, and potentially some water treatment plant parameters, in the development of the water quality sampling program.

- If test results are not routinely communicated to operators or others who make decisions based on distribution system water quality information, consider developing a communication plan and mechanisms for disseminating the information.

- Based on the water quality test results, if appropriate controls are not applied by distribution system operators, consider developing and implementing an appropriate set of controls that may be implemented based on the results of routine water quality testing.

- ▲ Any deviation of performance from established target ranges should result in rapid response to achieve and maintain the desired performance levels, as developed and documented by utility staff.
- If staff has not been involved in the development of the water quality control testing program or SOPs, or if SOPs have not been developed to support water quality testing procedures and response, including emergency response, consider the following steps:
  - ▲ Develop, review, or update SOPs for any required water quality control testing or response procedures, including emergency response procedures.
    - It is recommended to include all applicable staff in this process to help ensure buy-in and acceptance of distribution system goals and procedures across the organization.
    - Development of utility emergency response procedures should consider integration of the utility's water quality control policies, as well as a timeline for how frontline staff will handle emergencies, including notification of appropriate utility personnel and public health authorities, as required.
  - ▲ Depending on the current process control program, there may be a need for additional training. This training should focus on providing hands-on experience for all the frontline operators so that they have confidence to make needed system control adjustments.
  - ▲ Practicing responses to various situations builds confidence to act when faced with undesirable situations that require emergency response.

RECOGNIZING PERFORMANCE DEVIATIONS
Understanding

Water quality and operational system data can vary from day to day. Operators and managers must be able to detect deviations in performance, which are sometimes subtle, to take action before water quality is significantly affected. Although this skill is enhanced by experience, most operators can detect important deviations just by paying attention and looking carefully at the data.

There are several tools available to operational staff to help them recognize data trends (such as tabular data, statistics, and graphs). SCADA systems may be used by operators to view real-time data. Pressures, flows, tank elevations, pump status, and water quality sensor outputs are all examples of data that may be displayed on a SCADA system for operator use. Many systems allow for multiple data points to be displayed together, which can be helpful when attempting to correlate various data parameters. Some systems also have the ability integrate real-time data into their hydraulic model to provide a "live" update capability.

The volume of data available to operators can be overwhelming, particularly when considering online sensors that have the capability to produce a data point every second. Sometimes operators "lock" in on a few key data outputs that they are the most comfortable with and use them almost exclusively. This practice can be limiting and, in some cases, obscure important issues. Valuable and important data clues can be hiding in any data output and may also be recognized in the relationship between various distribution system data points. Operators need to have access to real time data trends that can quickly reveal significant performance deviations. As discussed later in this chapter, the application of appropriate system alarms and controls may also contribute to this process. The operator's critical role and responsibility in maintaining distribution system integrity is dependent on having data that is readily available, accurate, and understandable.

Noticing *change* is the key to assessing operational data, and, for this to be successfully accomplished, it is important that a performance baseline has been established, taking into account seasonal and other variations in performance that may regularly occur. Operators, water quality technicians, managers, engineers, and other responsible staff should be looking for unusual changes in data values, relative to the established baseline. Be aware, that sometimes *no change* is a change. If a normally highly variable parameter suddenly becomes steady, something may be wrong that requires additional investigation.

It is important that operations staff learn the details of the system. Staff needs to be constantly evaluating performance data to detect very slight variations from normal that may be indicative of an early warning to potential problems in the system. Attentive operators, through constant detailed observation, establish a baseline or benchmark that enables them to quickly identify and respond to deviations in the performance of critical system parameters. This is sometimes referred to as data-driven decision making.

How often data are reviewed, to detect a change, depends on the type of data and how quickly it may change. Pressure and flow are real-time data parameters that may change frequently. These data may need to be continuously monitored. The SCADA system can help because operators may set system alarms to alert staff when the when the value deviates beyond the established control range or changes beyond a preestablished amount. When an alarm occurs, it does not necessarily indicate an emergency, but it provides an indication to the operator that additional attention to the situation may be required. Some commercially available distribution system monitoring systems also have the ability to learn about the system's typical characteristics, alert operators about events that may occur, and attempt to identify their potential causes.

Some data may be reviewed daily or even less often; however, this does not mean that it is less important. Long-term trends can be indicators of very serious problems. For example, a slow decrease in chlorine residual at a distribution system location, over the period of several months, requires investigation to identify and address its root cause and prevent more serious problems from occurring.

Following are a few examples of assessing data using trend graphs. Operators should periodically review operational and water quality data to recognize performance deviations.

Chlorine Residual Example

In this example, the minimum and 95th percentile distribution system chlorine residual data are collected and reviewed. Additionally, these data are plotted on a trend chart monthly. Operational staff reviews these data to detect performance deviations. The performance goal for free chlorine residual is to maintain a minimum concentration of 0.20 mg/L. An example trend chart for chlorine residual concentration is displayed in Figure 5-1. The monthly average, 95th percentile, and minimum chlorine concentrations are plotted on this trend chart.

The 95th percentile value for August 2007 should be recognized by utility staff as a deviation. Although this value is higher than the utility's goal of maintaining a minimum concentration of 0.2 mg/L, it was the lowest 95th percentile concentration observed throughout the year. An event such as this should be investigated at the time it occurs to determine the primary cause of the low disinfectant residual. Operations staff may consider the location at which the low residual occurred to determine whether water age or low water usage contributed to the disinfectant residual concentration. That this low disinfectant residual concentration occurred in August should also prompt operations staff to consider any potential seasonal issues that may have contributed to the event, so that similar events may be able to be prevented in the future.

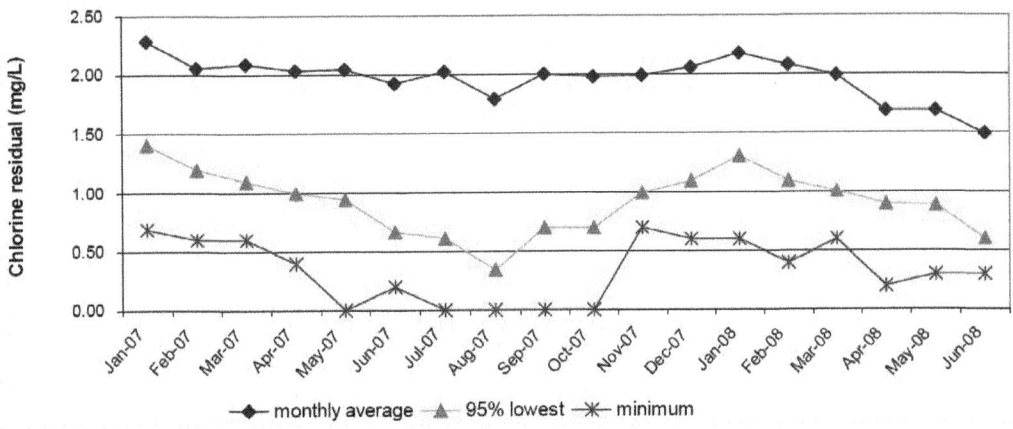

Figure 5-1. Example chlorine residual trend chart

Pressure Example

Pressure is often a real-time data output. When that is the case, the SCADA system continuously displays the values from pressure sensors at various locations. Pressure values can, in some cases, be very stable due to gravity system design; however, many systems can also experience a high degree of variability in pressure measurements, including transient pressure changes. When reviewing pressure data, operators should expand the pressure scale to enable them to see small variations in the pressure measurement. Data should also be collected from pressure sensors at a frequency that is sufficiently high for short-term changes in pressure to be recognized and recorded. Additional information about pressure monitoring is provided in chapter 2.

Figure 5-2 displays an example trend chart for pressure. This trend chart displays some degree of variability in pressure measurements. The pressure values on the chart also range from 158 to 161.3 psi, which may be high when compared with the average distribution system. For operators to evaluate the data and know whether additional evaluation and/or response is required, it is important for them to understand typical baseline conditions for the distribution system and each pressure zone. For example, if typical distribution system pressures range from 80 to 90 psi, then the cause of high-pressure readings displayed on the trend chart for this location may require further investigation.

Main Break Example

The frequency of main breaks is a measure that is not always reviewed by system operators. However, operators should become more aware of the value of main break data because they provide multiple insights into distribution piping integrity (recall that main break frequency serves as an indicator of the system's

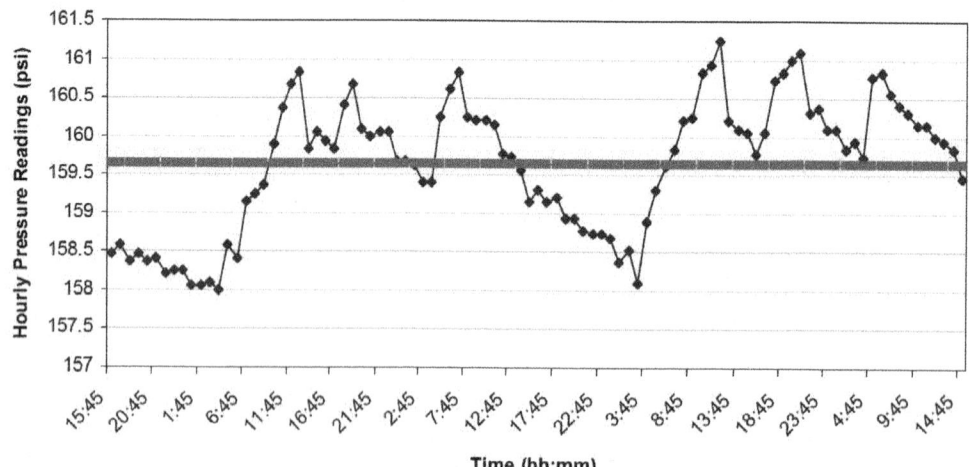

Figure 5-2. Example pressure trend chart

physical integrity) and, potentially overall system operations. Operators need to recognize that their process control decisions can influence these criteria over time. It may be necessary to optimize several variables (including internal and external corrosion control, pressure stability, material selection, pipeline replacement, and rehabilitation) to slowly change this measure. Several of these variables are available to operators and should be considered. Additional information about main breaks is included in chapter 2.

Figure 5-3 displays an example trend chart for main break frequency. In contrast to the previous figures, in which data were plotted on a daily or monthly basis, main break frequency is commonly considered on an annual basis to evaluate the utility's year-to-year progress. This chart compares the utility's data to the PSW optimization goal of <15 reported leaks or breaks per 100 miles of utility-owned pipeline on an annual basis. Although the utility data do not meet the Partnership's optimization goals for main breaks for every year that data is plotted, the general downward trend in main break frequency over time indicates that progress is being made toward optimization.

Status

The status of staff's ability to recognize performance deviations, as it relates to achieving a desired level of distribution system performance, may be assessed by reviewing the following self-assessment questions. The self-assessment team is not limited to these questions and may consider discussing additional topics

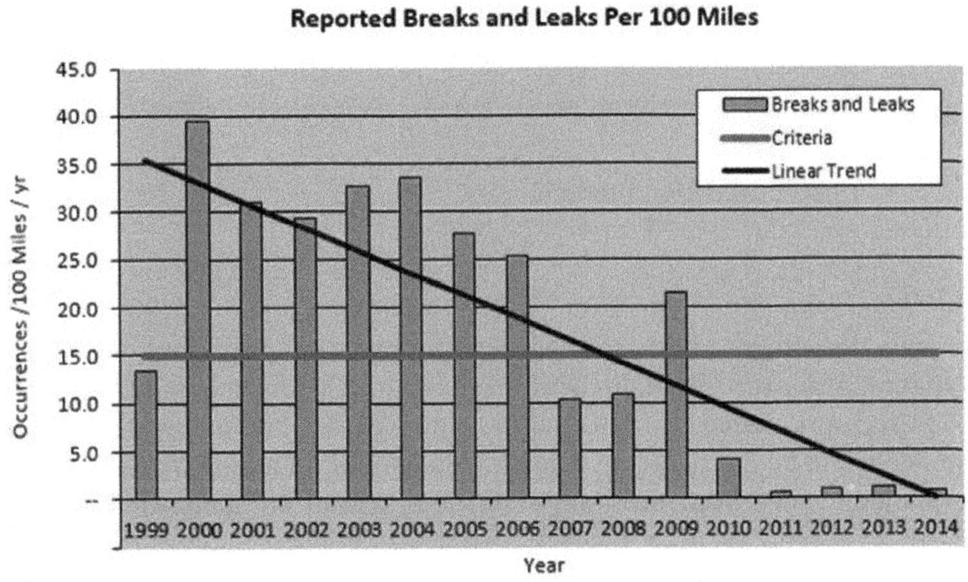

Figure 5-3. Example main break frequency trend chart

related to recognizing performance deviations that may assist in identifying and addressing performance-limiting factors.

- Does operations staff review data frequently enough to maintain water quality within optimization ranges established and recognize performance deviations?

  ▲ In general, the more variable the data, the more frequent the interval at which that data should be collected, graphed, and reviewed. When addressing this question, consider the data review frequency for the following parameters:

    ▪ Pressure and flow

    ▪ Disinfectant residual

    ▪ Main break frequency

    ▪ Water quality data

    ▪ Customer complaints

- Are spreadsheets, tables, or other tools used by operators to turn data into information that can be used to make process control decisions? Are data summarized with charts and tables, available to all appropriate staff for review? Is a distribution system team assembled to review data periodically?

- Do operators receive training on data review and analysis that includes the following topics?

  ▲ Access to appropriate data outputs

  ▲ Use real data to recognize changes that may require investigation

  ▲ Highlight any situations in which water quality deterioration was averted

  ▲ Data invalidation protocol

Action

If areas of the utility's ability to recognize performance deviations are considered to have a status of Partially Optimized or Not Optimized, and this parameter has been prioritized and selected for improvement, the following steps are recommended to be considered in the development of an action plan to improve performance in this area, including implementation of an appropriate training program, if needed.

- If staff does not review performance data frequently enough, as required by the variability of the data, to maintain water quality within optimization ranges established, a program should be developed to organize the evaluation process. Such a program should include representatives from all affected departments.

- If spreadsheets, tables, or other tools are not used by operators to turn data into information that can be used to make operational control decisions, develop a system to accomplish this.

  - ▲ It is recommended that such a system include a visual means for data evaluation, such as trend charts, and that a mechanism is in place to communicate the information to all appropriate staff members.

  - ▲ Consider encouraging operations staff to meet on a regular basis to review the data, learn about baseline patterns and identify performance deviations, and determine the root cause for any performance deviations that may occur.

- If staff does not receive annual training covering data evaluation and review, develop an appropriate training program, that includes all data evaluation and review procedures that the utility may use.

  - ▲ Consider including utility case studies in such training, which can demonstrate appropriate operator responses that successfully averted water quality deterioration as well as those that may provide examples of lessons learned.

SYSTEM EVALUATION RESPONSE

Understanding

Improvement of system performance is ultimately achieved by implementing control procedures that can be used to move a capable system to achieve the desired water quality goal (see Figure 1-1). Successful system control involves delivering consistent, high-quality water through a complex distribution network. To achieve this, distribution system staff must be able to use a variety of system controls and understand the implications of their actions.

It is especially important that distribution system staff operating newer distribution systems, or those that are less complex in nature, takes measures to avoid complacency and have the ability to maintain operational control skills. Complacency can inhibit response to rapidly changing conditions and can result in a loss of control of system performance.

Although distribution system control is the most important activity required to achieve consistent, reliable distribution system performance, distribution system operators must also understand and apply other concepts. For example, it is important for distribution system operators to also understand the water treatment plant processes because of their potential impact on distribution system performance. Distribution system and treatment plant operators both must also understand water chemistry and microbiology, because changes in either have the potential to cause water quality deterioration. Beyond understanding and applying technical concepts of water treatment, staff must also have the tenacity to achieve operational excellence and the confidence to make the required operational changes. The confidence to change an existing practice empowers the staff to respond appropriately to variable conditions.

Status

The status of the utility's system evaluation response, as it relates to achieving a desired level of distribution system performance maybe assessed by reviewing the following self-assessment questions. The self-assessment team is not limited to these questions and may consider discussing additional topics related to system evaluation response that may assist in identifying and addressing performance-limiting factors.

- Does all operating staff have the following characteristics?
  - ▲ The tenacity to achieve the performance goals. When system performance starts to deteriorate, does all of the staff actively participate in finding potential causes and implementing solutions until performance improves?

  - ▲ A willingness to take responsibility for system performance. When system performance starts to deteriorate, does the frontline operations staff (regardless of the department) assume responsibility to identify potential causes and correct the problem, or do they rely on supervisors, or depend on other departments, to tell them what to do?

  - ▲ A willingness to learn. Is the staff interested and willing to learn new ideas and control procedures that support the system performance, or is there an attitude that current practices are acceptable and nothing needs to change?

  - ▲ Confidence to make changes. Does the staff have the curiosity and capability to investigate alternative strategies (for example, change flow patterns) and the confidence to change an existing practice to optimize performance?

  - ▲ Empowerment to make changes. Does the staff feel that it has the support of management necessary to make needed control adjustments, or are actions limited by a perception that they do not have the proper authority?

  - ▲ Understanding of when to call for help and whom to call. When performance continues to deteriorate after the staff has made changes, are they willing to call for assistance and do they know where to call for that assistance?

- Is the operating staff able to communicate their concern regarding water quality deterioration, in particular the disinfectant residual level, to other departments? Consider the following factors in this response:
  - When a low disinfectant residual is identified, can this be communicated to departments that have the ability to take action?
  - Will the department contacted to respond understand the issue and react with urgency?
  - After appropriate action to correct the situation has been taken, is there follow-up testing to determine if the action is successful? Is this information communicated to all parties involved?
  - Are SOPs and protocols developed and implemented to ensure a consistent, appropriate, and timely response?
  - Has an emergency response procedure been developed for loss of or low disinfectant residual levels or to respond to unacceptable water quality conditions, when detected?
  - Is routine training provided on appropriate responses to water quality related issues?
- Does utility staff, working as a team, practice "what-if" thinking to develop contingency plans for responding to abnormal or atypical distribution system situations?
- Can the appropriate operating staff conduct the necessary hydraulic model simulations to determine the most appropriate response to a given situation—or does the staff have the ability to contact a party that does?
  - Is a calibrated working version of the system hydraulic model available to designated system operators 24/7?
  - Do system operators have the training to use the model to help understand the impact of an emergency situation?
  - If operators do not have the required expertise to run the model, is an expert available that can provide 24/7 response?
- Are water quality data shared with system operators so that they can make operational adjustments that result in improvement?
  - Operators must be trained in the evaluation of water quality information.
  - The implications of disinfectant residual levels, bacteria test results, and DBP concentrations should be well understood.
  - SOPs and protocols should be developed and implemented to help ensure a consistent, appropriate, and timely response.

- Does appropriate staff have the knowledge to maintain process components (pumps, valves, etc) and understand the tasks of a preventive maintenance program?

- Is the distribution system staff evaluating appropriate parameters, including daily demand, to help determine and control water age?

  - ▲ A system hydraulic model should have the ability to calculate water age at key locations throughout the system.

  - ▲ Operators should be trained and have the capability to make adjustments to move water strategically in the system to reduce water age.

  - ▲ System staff may need to use flushing to remove water that has deteriorated due to excessive water age. Flushing SOPs should include guidance on how long to flush and how to evaluate the effectiveness of flushing. Refer to chapter 3 for additional information.

  - ▲ When flushing is used, field personnel must understand how to apply the procedure effectively to ensure that the poor quality water is not distributed through the system.

  - ▲ Appropriate water quality testing should be used to determine when to terminate a flushing procedure.

- Do operators review their procedures to ensure that long-term performance objectives are included? Consider the following:

  - ▲ Procedures involving pressure levels and control may influence main break frequency and backflow occurrence risk.

  - ▲ Disinfectant residual concentrations can impact DBP formation.

  - ▲ Customer complaints may be an indicator of developing problems.

  - ▲ Corrosion control efforts may help to reduce main break frequency and improve water quality.

Action

If areas of the utility's system evaluation response procedures are considered to have a status of Partially Optimized or Not Optimized, and this parameter has been prioritized and selected for improvement, the following steps are recommended to be considered in the development of an action plan to improve performance in this area.

- If staff does not show tenacity for supplying high-quality water on a continuous basis, consider ways to create an environment conducive to developing a staff with these characteristics. Make sure that any training plans developed to support this area ensure management support of performance goals and empowering frontline operators to make operational control adjustments to maintain distribution system performance.

- If the potential exists for staff to become complacent in their activities, which may have a greater potential for staff operating newer or less complex systems, implement strategies to combat complacency, including training, tabletop exercises, and encouraging proactive practices, such as the development of contingency plans.

- If staff does not have access to a calibrated hydraulic model, 24/7, or the ability to contact an individual or entity that does, a program should be developed to make this available. If staff will be using the model, provide training in its use and procedures to calculate water age.

- If staff understanding of preventive maintenance programs and procedures is limited, take measures to build knowledge, expertise, and procedures relating to preventive maintenance of distribution system components.

- If staff does not regularly receive information regarding water quality or does not understand the consequences of such information, a training program should be developed that covers this area. The program should include relevant science and operational principles. It may be advantageous to include staff from various departments that need to work together to respond to water quality situations.

COMMUNICATION

Understanding

Effective communication is the cornerstone of any successful organization. A utility should place high priority on the dissemination of information to ensure it is properly conveyed and comprehended throughout all levels of operations to maintain optimized system performance. Communication flows in a multitude of directions; including horizontally, from peer to peer, vertically from management down, as well as from frontline up within the organizational structure.

Communication is usually classified as either formal or informal. Formal communications are most commonly recognized as informational or instructional and take place within a framework of the utility's established administrative structure. This type of information transfer may be accomplished through verbal means, such as through meetings with a group or individual, or written as in policies, standard operating procedures, posted notices, e-mails, or electronic media. Informal communication may contain elements of formal communication but is generally recognized as bringing about the greatest teamwork and is the most common form used in day-to-day operations. The effectiveness of informal communication is essential, because many workforces depend on this type of information transfer between employees to accomplish system tasks, such as working safely, maintaining optimized water quality, evaluating potential system issues, and reviewing equipment operational status and maintenance requirements.

An often overlooked component of communications is a check for comprehension. Most organizations are aware that it is important to provide information and instruction to their employees, as well as arrange a means for the employees to voice their needs and concerns, but many overlook that it is just as important to check for understanding. Misunderstandings and differing interpretations among individuals often lead to unintended consequences. Accurate comprehension will help ensure that actions are taken in accordance with given directives, procedures, and established goals.

Just as there are procedures for plant operation and maintenance, having similar communication procedures can benefit distribution system operations. For example, developing procedures that allow system staff to better understand who to contact, under what circumstances or performance triggers, and in what format can help promote proper communications when they are needed, resulting in a rapid response to system issues. The establishment of communication forums, such as "tailgate talks" or operations meetings can also help to promote expanded communications and information sharing in other areas.

Communication at a water utility has many potential challenges; different shifts, workdays and locational separation, and levels of understanding; however, with perseverance and tenacity, these obstacles can be overcome. Use all appropriate modes of communication such as e-mail, comprehensive log entries, face-to-face meetings, and possibly electronic blogs or websites to disseminate information that contributes to successful system operation.

Status

The status of the utility's communication practices, as they relate to achieving a desired level of distribution system performance maybe assessed by reviewing the following self-assessment questions. The self-assessment team is not limited to these questions and may consider discussing additional topics related to communications that may assist in identifying and addressing performance-limiting factors.

- Has a formal communication protocol for all critical distribution system staff been developed?
  - ▲ Meetings should be scheduled at regular intervals with planned agenda.
  - ▲ A common area for posting notices and information should be considered.

▲ Disseminated information should be readily available to employees by a various means such as e-mail accounts or searchable intranet sites.

- Do system operators from various shifts and disciplines effectively communicate with each other?

  ▲ There should be a deliberate exchange amongst system operators to ensure pertinent operational information is passed along from one group to another.

  ▲ It is suggested that a designated time exists for operators to meet with each other with the system supervisor. These may consist of daily, weekly, or monthly meetings. These meetings should be scheduled in a matter that is compatible with the work shift schedule.

- Do system operators and maintenance workers effectively communicate to ensure distribution system maintenance status is provided to all system operators frequently and accurately?

  ▲ One of the most vulnerable areas of system performance is when atypical system repairs or operations are taking place. It is critical that maintenance status is communicated to operators in a timely, efficient manner.

- Is information communication encouraged between treatment plant staff and distribution system staff?

  ▲ The utility should encourage positive employee relationships to facilitate the exchange of information.

  ▲ Operational status and work in progress should be communicated between departments and employees through written or verbal means. This may include log sheets, data collection spreadsheets or graphs, status boards, or briefing sessions with individuals or groups.

- Does the utility ensure that employees understand the information presented?

  ▲ Administration and management should not assume that the employees have complete comprehension without seeking confirmation. Always allow for questions and, if appropriate, request the employee to restate the conveyed material.

  ▲ Staff should be skilled in communication techniques to recognize verbal and nonverbal cues that confirm acceptance and comprehension.

- Does utility leadership ensure that subordinate employees receive and understand the information in a timely manner, especially distribution system operators who spend the majority of their time in the field?

Action

If areas of the utility's communication practices are considered to have a status of Partially Optimized or Not Optimized, and this parameter has been prioritized and selected for improvement, the following steps are recommended to be considered in the development of an action plan to improve performance in this area.

- Participating in training programs to enhance communication skills.
- Evaluating meeting structure for effective leadership, content, and audience reception.
- Facilitating opportunities or methods for information transfer between employees.
- Developing a culture to encourage employee questions and inquiries and providing opportunities for employees to share issues and concerns.
- Systematically reviewing written communications for validity and updating requirements.
- Providing a structured system for information transfer. Examples of this include, but are not limited to, log books, data spreadsheets or graphs, bulletin boards, memos, e-mails, tailgate sessions, and one-on-one meetings.
- Implementing a process to verify communication comprehension.

ONLINE INSTRUMENTATION/SCADA

Understanding

The use of SCADA is essential to address control and reliability from the source water, through the water treatment process, and to the ends of the distribution system. SCADA enables system operators to view live and historical data, trend critical parameters, compare parameters, and create reports. The information provided by the SCADA system is a critical tool that staff uses to help document distribution system performance and make appropriate system decisions.

In this section, all field devices and any equipment controlled or parameters measured by staff will be referred to as *instrumentation*. Staff uses instrumentation to generate data that helps them manage the distribution system. Instrumentation and the associated data historian capabilities should be compliant with local regulatory requirements as well as the requirements of system staff. The decision to fully optimize, control, and manage SCADA systems efficiently will require input from all parties, including the operational staff.

Instrumentation and a SCADA system are tools in the system operator's toolbox that can be used to learn about a distribution system and control system operations. Information from the SCADA system should not merely be collected and stored but should be evaluated and communicated to system staff.

This can be achieved using a variety of methods, as described in the previous section, including tailgate meetings, shift change meetings, operational meetings, or electronically. Regularly examining and evaluating data generated by instrumentation allows staff to learn more about operations, develop changes that can improve performance, and develop control software settings that allow for improved operational control.

Instrumentation can alert the system operator and serve as an early warning that operations may need to change due to demand changes based on weather conditions or changes caused by a manmade event, helping system operators to avoid complacency. For example, one indicator that changes in system operation may be required occurs in the case where a chlorine analyzer, monitoring water leaving a storage tank, reveals a drop in chlorine residual. As the chlorine residual begins to approach an action level, the system operator will follow an SOP to make an adjustment and correct this event. In some cases, the instrument's signal may also be used to signal the control system to initiate the appropriate response processes, or at minimum generate an alarm that can help to serve as a "safety net" for maintain optimized distribution system water quality.

Although the complexity of control systems may vary from utility to utility, the benefits that SCADA control can bring to the distribution system of are independent of system size. Regardless of size or configuration, utilities may wish to consider the use of instrumentation and SCADA to collect appropriate distribution system information and apply controls that are critical to maintaining water quality. These are tools that can also aid utilities in completing the self-assessment process and achieving optimization.

The operator of a distribution system may be compared with a detective, proactively looking for the anomalies in the system rather than maintaining a reactive mentality to system changes. Well-maintained and reliable instrumentation helps the system operator make timely decisions that lead to optimized system performance on a consistent basis. Making the decision to add instrumentation may require planning for purchase or replacement, including redundant instrumentation and spare parts. As utilities develop action plans for improvement, consider that the addition of instrumentation, and the equipment they control, may require long-term budgeting, a step that should be taken into account when developing action items associated with instrumentation. Budgeting consideration should take into account any costs associated with installation and ongoing equipment maintenance.

Proper calibration, verification, and maintenance are required to ensure accuracy and reliable performance of instrumentation. It is important to follow the manufacturer's instructions for the type of instruments in use in the system. If online instrumentation is not reliable and not producing accurate data, this can lead to poor decision-making, potentially putting the distribution system and water quality at risk. Operational staff must have confidence in their instrumentation so that they are able to make well-informed system decisions.

Online instrumentation is often used in conjunction with grab sampling, although one does not necessarily take the place of the other. There are often minimum frequencies at which an operator is required to perform laboratory analyses. For example, chloraminated utilities collect total chlorine, monochloramine, and free ammonia paired grab samples at specific locations throughout the distribution system on a regular basis. These results are recorded for evaluation and comparison with the on-line instrumentation located at the entry point to the distribution system. A well-maintained data historian should be able to capture readings from instruments, archive data, and compress the data to 15-minute intervals, or intervals selected by the utility, without gaps in the data collection. Maintenance records should also be kept in hard copy and/or electronic form. Currently, data can also be stored on a secure, hosted site. Utilities must remember that the they are responsible for data collection, archiving, and ensuring the integrity of the data, regardless of where those data are stored. Data gaps may need to be accounted for by testing in the laboratory, which may be a regulatory requirement. If online instrumentation or its associated data collection system is down, the results of laboratory tests may need to be collected at certain intervals to ensure system control and regulatory compliance until the problem is resolved.

A regulatory audit may require the provision of documentation and records of calibration, verification, and maintenance to the regulating authority for online instruments in use. In the laboratory, logs are also to be maintained for calibration, verification, and maintenance of devices to verify they operate in an acceptable manner as prescribed by the manufacturer. Where appropriate, spreadsheets should be developed and updated annually for each piece of analytical equipment.

In a smaller system, calibrations and verifications may be performed by a licensed distribution system operator, or outside calibration and maintenance contractor working closely with utility staff, who has a comprehensive understanding of instrumentation. In a larger facility, the calibration technician or specialist who works on instrumentation may not necessarily be a licensed operator; however, the person performing these functions should have an understanding of the importance of the instrumentation and its role in maintaining water quality. The individual who performs the calibrations and verifications may be site-specific, but the main objective of the process is to ensure instrumentation is operating in an acceptable range and providing accurate data to SCADA and the data historian.

The instruments and devices used by the utility may be located in a variety of locations, some of which may be a significant distance from the central computer(s). Communications may involve travel of the signal through an elaborate process from the field. When completing the self-assessment, the ability for staff to understand the network in use is important. Operators should have a basic understanding of how the instruments send signals to SCADA and how control devices react to the signals, although expert knowledge of the SCADA system, programming, and advanced troubleshooting is not required to use the valuable information that SCADA can provide. As stated previously, the system operators must determine what data to collect and the most suitable frequency of data collection.

Status

The status of instrumentation is extremely important for completing the self-assessment, communicating results to regulatory agencies, and for operational staff to maintain a proactive approach. Each water quality parameter addressed in this guide requires different instrumentation and data collection requirements for distribution system operators to maintain a confidence level with the water quality and the analyzers used to monitor it.

The status of instrumentation, as it relates to achieving a desired level of distribution system performance maybe assessed by reviewing the following self-assessment questions. The self-assessment team is not limited to these questions and may consider discussing additional topics related to instrumentation that may assist in identifying and addressing performance limiting factors.

- Has a maintenance and calibration schedule been implemented for all instrumentation?
  - ▲ Appropriately prioritized maintenance is required to ensure reliability of the instrumentation.
- Has the utility implemented a training program to train system operators or maintenance technicians in the care and maintenance of on-line water quality instrumentation? Has the utility implemented a training program to train system operators in the use and operation of the SCADA system and field SCADA interfaces? (An example of an operator performing instrument maintenance is shown in Figure 5-4)
  - ▲ Utilities have also reported that maintaining a contract for calibration and maintenance services with an outside provider that works closely with utility staff can be useful in keeping online water quality instrumentation operating accurately.
- Does the utility have clearly defined roles on who is permitted to modify and maintain the SCADA system and are those individuals properly trained?
- Application and placement—are the instruments selected suitable for the application and physical placement for which they are installed? Does the placement of the instruments allow for ease of access for maintenance? Are environmental conditions suitable for analyzer placement? Is the placement of the sample collection and delivery lines suitable to obtain a representative sample?

Source: Aurora Water

Figure 5-4. Operators should be trained on instrument maintenance procedures to help ensure data accuracy

- Calibration/verification—are the instruments calibrated and verified regularly? Do system operators follow the manufacturer's instructions for calibration and verification? Are calibration and verification performed at a frequency that, at minimum, meets regulatory requirements? Are calibration records maintained and accessible when needed? Has an SOP been developed for the calibration and maintenance of all critical instrumentation?

- Are control systems tested and checked regularly to ensure proper operation and functionality when needed? This may include testing functionality of alarms and control capabilities.

- Redundancy—are redundant instruments and equipment installed where appropriate? Does the system have a plan established for the procurement of all critical parts to move toward making the system "bulletproof"?

- Does the utility have provision for a backup server or SCADA system in the event of communication loss or failure? Does the distribution system operator have adequate quantity of equipment to monitor SCADA from the field? Is there an SOP to manually operate the distribution system in the absence of SCADA or a prolonged power outage?

- Is all pertinent information documented? Is record-keeping adequate (i.e., is the necessary information maintained for the amount of time required by regulatory parties or internal system processes)? Ensure

SOPs, logbooks, instrument manuals, and work performed such as calibration or verification is completed and archived.

- Do distribution system operators have adequate capability to view critical system data while working in the field?

- Does the control system provide operational "safety nets," consistent with system practices and performance goals, to protect the system and water quality? For example, a steady decrease of chlorine residual over an extended period could trigger unidirectional flushing in the affected portion of the distribution system.

Action

If areas of the utility's instrumentation and SCADA system are considered to have a status of Partially Optimized or Not Optimized, and this parameter has been prioritized and selected for improvement, the following steps are recommended to be considered in the development of an action plan to improve performance in this area.

- If adequate maintenance and calibration are not being performed on instrumentation, develop a formalized and prioritized schedule for completing this work, including identifying the responsible parties. Ensure that all staff is trained in the appropriate calibration and maintenance tasks and consider developing SOPs for these items.

- To ensure operators are trained and confident in troubleshooting and operation of online instrumentation, develop a training program and SOPs using internal and external subject matter experts as appropriate to increase instrumentation knowledge.

- If instrumentation performance and maintenance is being affected by poor instrument selection or placement, select a more appropriate instrument or installation site.

  ▲ The distance between the sample and the analyzer should be minimized to prevent lag time in analysis.

  ▲ Instrument selection should be based on the sample type and expected concentration range for the parameter of interest, taking into account any additional issues, such as the potential for waste generation and disposal.

  ▲ Instruments should be installed in locations that are accessible for calibration, verification, and maintenance activities.

  ▲ Good judgment calls should be made when upgrading instrumentation to select the type of analyzer that is most suitable for the system's application and intended use. Technical information provided by the instrument manufacturer may help with this process.

- To help ensure data validity, establish a procedure for instrument and data verification. This may include verification through the use of comparative grab sampling or purchased or prepared standards.

  ▲ When the results of online analyzers are compared with grab samples, utilities will typically define an acceptable percent deviation for agreement (for example, 10%). This acceptable range may be parameter, instrument, and sample specific. If results do not agree within the acceptable deviation range, further troubleshooting, calibration, or maintenance activities may be required. An example grab sample comparison log for chlorine is displayed in Figure 5-5.

- If insufficient spare parts are available onsite for critical analyzers, consider developing a plan to budget for and procure the required materials. Consider purchasing redundant instrumentation and/or equipment for critical parameters if this is required to achieve the system's performance goals.

- In cases where record-keeping is insufficient or inconsistent, establish procedures for recording, maintaining, and storing records related to instrumentation. Ensure that this information is available to all staff and that staff is familiar with data recording procedures and data access locations.

  ▲ If records are stored in a variety of separate locations, consider integrating the information to allow information to be accessed and correlated more rapidly. Regulatory requirements may exist for data

Method 334.0 Routine Comparison Calibration for Online Chlorine Analyzers

PWSID#: _____   Plant Name: _____   Month/Year: _____

Online Analyzer Make/Model #: _____   Online Analyzer Monitoring Location: _____

Grab Sample Colorimeter Make/Model #: _____   Date of Last Colorimeter QA/QC: _____

Date /Time of Comparison Check	Initials of Operator Conducting Comparison Check	(A) Grab Sample Result via Verified Colorimeter Mg/L Free Cl2	(B) Online Analyzer Reading at Time of Grab Sample Mg/L Free Cl2	(C) Calculate % Difference ((A-B) ÷ A) X100	(D) If Column (C) is greater than (>) 15%, Summarize actions taken in this column. Options include: (1) Conduct Online Analyzer QA/QC troubleshooting & perform another comparison check* (2) If Online Amperometric Analyzer, adjust reading to match DPD Grab Sample Result as per instrument manual (3) If Online Colorimetric analyzer, contact the instrument manufacturer for calibration guidance *Contact instrument manufacturer for QA/QC Guidance. Be sure to complete and record second comparison check after any equipment calibration or adjustments

Figure 5-5.   Example grab sample comparison log for entry point chlorine analyzer results, using USEPA Method 334 (courtesy of Pennsylvania Department of Environmental Protection PA DEP)

storage and record-keeping, and it is important that system staff be familiar with these requirements.

- Ensure that the control system provides "safety nets" to protect water quality. Input from staff will be an important consideration in developing these controls. Be sure that the levels used in system control programming are consistent with the system's standard operating procedures. Consider the use of multiple levels of controls or alarms, such as using a "high" alarm in conjunction with a "high-high" alarm, each with different associated consequences and control actions.

Instrumentation and control systems are important tools that can be used to improve operations, system performance, and water quality. There is a wide variety of actions that staff may take to improve the way that instrumentation is used and applied and how the data produced is collected, stored, evaluated, and communicated. Although the self-assessment focuses on the most critical aspects of instrument use and those most applicable to a broad segment of distribution systems, utilities are encouraged to explore areas outside of these to develop an action plan to improve instrumentation performance in the most relevant areas. Even the most robust SCADA system is simply an aid to the system operator in delivering the highest quality water possible. The information that SCADA collects, and the processes it controls, aid the operator, but the system operator is still responsible for verifying the SCADA information by frequently checking equipment function, laboratory analyses, and other safeguards, a process that can be accomplished through regular data analysis and system evaluations.

Performance Limiting Factors Summary

Table 5-3 summarizes factors related to operational control that may be limiting optimized performance of the distribution system. Check whether these factors are Optimized and Documented, Partially Optimized, or Not Optimized. Factors identified as Partially Optimized or Not Optimized will be prioritized in chapter 7, Identification and Prioritization of Performance Limiting Factors, where an action plan can be developed and implemented that allows for optimization of parameters selected for improvement.

Table 5-3.   Application of operational concepts

Self-assessment category	Questions for gauging optimization status	Optimization status			Comments
		Optimized and documented	Partially optimized	Not optimized	
Water Quality Control Testing	Does the utility have a routine distribution system water quality testing program, including established water quality goals and sampling schedules?				
	Are test results communicated to distribution system operators?				
	Do operators make adjustments to the appropriate system controls based on the water quality test results?				
	Has all of the operational staff been involved in the development of the system control program and SOPs, including emergency response guidelines?				
Recognizing Performance Deviations	Does operations staff review data frequently enough to maintain water quality within optimization ranges established and recognize performance deviations?				
	Are spreadsheets, tables, or other tools used by operators to turn data into information that can be used to make process control decisions? Are data summarized with charts and tables, available to all appropriate staff for review? Is a distribution system team assembled to review data periodically?				
	Do operators receive training on data review and analysis that includes the following topics? • Access to appropriate data outputs • Use real data to recognize changes that may require investigation • Highlight any situations in which water quality deterioration was averted • Data invalidation protocol.				

Self-assessment category	Questions for gauging optimization status	Optimization status			Comments
		Optimized and documented	Partially optimized	Not optimized	
System Evaluation Response	Does all operating staff have the following characteristics? • The tenacity to achieve the performance goals • A willingness to take responsibility for system performance. • A willingness to learn. • Confidence to make changes. • Empowerment to make changes. • Understanding of when to call for help and whom to call.				
	Is the operating staff able to communicate their concern regarding water quality deterioration, in particular the disinfectant residual level, to other departments?				
	Does utility staff, working as a team, practice "what-if" thinking to develop contingency plans for responding to abnormal or atypical distribution system situations?				
	Can the appropriate operating staff conduct the necessary hydraulic model simulations to determine the most appropriate response to a given situation—or does the staff have the ability to contact a party that does?				
	Are water quality data shared with system operators so that they can make operational adjustments that result in improvement?				
	Does appropriate staff have the knowledge to maintain process components (pumps, valves etc) and understand the tasks of a preventive maintenance program?				
	Is the distribution system staff evaluating appropriate parameters, including daily demand, to help determine and control water age?				
	Do operators review their procedures to ensure that long-term performance objectives are included?				

Self-assessment category	Questions for gauging optimization status	Optimization status			Comments
		Optimized and documented	Partially optimized	Not optimized	
Communication	Has a formal communication protocol for all critical distribution system staff been developed?				
	Do system operators from various shifts and disciplines effectively communicate with each other?				
	Do system operators and maintenance workers effectively communicate to ensure distribution system maintenance status is provided to all system operators frequently and accurately?				
	Is information communication encouraged between treatment plant staff and distribution system staff?				
	Does the utility ensure that employees understand the information presented?				
	Does utility leadership ensure that subordinate employees receive and understand the information in a timely manner, especially distribution system operators who spend the majority of their time in the field?				
Online Instrumentation and SCADA	Has a maintenance and calibration schedule been implemented for all instrumentation?				
	Has the utility implemented a training program to train system operators or maintenance technicians in the care and maintenance of on-line water quality instrumentation? Has the utility implemented a training program to train system operators in the use and operation of the SCADA system and field SCADA interfaces?				
	Does the utility have clearly defined roles on who is permitted to modify and maintain the SCADA system and are those individuals properly trained?				

Self-assessment category	Questions for gauging optimization status	Optimization status			Comments
		Optimized and documented	Partially optimized	Not optimized	
Online Instrumentation and SCADA (*continued*)	Application and placement: Are the instruments selected suitable for the application and physical placement for which they are installed? Does the placement of the instruments allow for ease of access for maintenance? Are environmental conditions suitable for analyzer placement? Is the placement of the sample collection and delivery lines suitable to obtain a representative sample?				
	Calibration/verification—are the instruments calibrated and verified regularly? Do system operators follow the manufacturer's instructions for calibration and verification? Are calibration and verification performed at a frequency that, at minimum, meets regulatory requirements? Are calibration records maintained and accessible when needed? Has an SOP been developed for the calibration and maintenance of all critical instrumentation?				
	Are control systems tested and checked regularly to ensure proper operation and functionality when needed? This may include testing functionality of alarms and control capabilities.				
	Redundancy—are redundant instruments and equipment installed where appropriate? Does the system have a plan established for the procurement of all critical parts to move toward making the distribution system "bulletproof"?				

Self-assessment category	Questions for gauging optimization status	Optimization status			Comments
		Optimized and documented	Partially optimized	Not optimized	
Online Instrumentation and SCADA (continued)	Does the utility have provision for a backup server or SCADA system in the event of communication loss or failure? Does the distribution system operator have adequate quantity of equipment to monitor SCADA from the field? Is there an SOP to manually operate the distribution system in the absence of SCADA or a prolonged power outage?				
	Is all pertinent information documented? Is record-keeping adequate (i.e., is the necessary information maintained for the amount of time required by regulatory parties or internal system processes)?				
	Do distribution system operators have adequate capability to view critical system data while working in the field?				
	Does the control system provide operational "safety nets," consistent with system practices and process performance goals, to protect the operator and water quality?				

Reference

AWWA, 2001. M2—*Instrumentation and Control*, 3rd ed. Denver, Colo.: AWWA.

CHAPTER 6

ADMINISTRATION

Administrative practices can significantly impact a utility's ability to optimize operations and performance. As a result, all distribution systems completing the self-assessment process, regardless of configuration, should address the assessment questions in this chapter to identify performance-limiting factors and develop an action plan for improvement. With performance-limiting factors identified, a utility is then able to prioritize them on many different levels. For example, which ones will aid utility staff in facilitating "buy-in" of the process? Which performance-limiting factor corrections will garner the greatest improvement for the least investment? Which performance-limiting factor corrections will be the most capital intensive and require the greatest planning to correct? These questions and more are necessary for the utility to optimize its distribution system. Although it is important for the utility assessment team to agree upon the goal, there is a multitude of successful ways to get there.

The focus of this chapter is the assessment of administrative factors relative to overall utility performance, with a special focus on the distribution system. The administrative factors included in the self-assessment are listed next. This chapter is divided into sections that address each of these factors.

- Administrative Policies
- Acceptance of Optimization Goals
- Involvement of all Parties in the Partnership Process
- Documentation/Demonstration of Addressing Complacency
- Training
- Staffing
- Funding

Evaluation of administrative performance-limiting factors is subjective and based on management and staff involvement as part of the utility's self-assessment team. As described in previous chapters, the self-assessment team optimally should incorporate personnel from all disciplines and levels of the organization, including management. With management involvement throughout the self-assessment process, the questions and topics presented in this chapter should be addressed in an open fashion and serve to promote communication throughout the organization. Budgeting and financial planning, staffing levels, and administrative policies are the mechanisms that distribution system owners/administrators generally use to implement utility objectives. A fully optimized system will demonstrate a free flow of information throughout the organization. A potential sign of complacency is identifying all administration factors as optimized when system performance is not. A single table of references for this chapter is provided in Table 6-3, at the end of the chapter.

Partnership for Safe Water and Effective Utility Management

The Partnership for Safe Water (PSW) guidance and self-assessment process is designed to be complementary and compatible with existing resources and programs for water utility operations and management. One of these resources is AWWA's Utility Management Standards, which include AWWA G200-15: Distribution Systems Operation and Management. This standard includes detailed descriptions of the essential or critical requirements for the effective operation and management of a water distribution system. Many of these principles are echoed throughout the Partnership's distribution system self-assessment process.

Effective Utility Management (EUM) resources are other tools that drinking water utilities may apply to assess and continuously improve utility management practices. EUM is built around the Ten Attributes of Effectively Managed Utilities and Five Keys to Management Success, which cover all aspects of utility operations and is a well-accepted framework to help utilities improve their performance and move toward sustainable operations. EUM resources were developed by a collaborative partnership of several agencies, including the Association of Metropolitan Water Agencies, the American Public Works Association, AWWA, the National Association of Clean Water Agencies, the National Association of Water Companies, and the United States Environmental Protection Agency. Many of these organizations also participate in Partnership programs, such as the Partnership for Safe Water and Partnership for Clean Water. The Ten Attributes include areas such as Operational Optimization, Product Quality, Operational Resiliency, Financial Viability, Employee Leadership and Development, and Stakeholder Understanding and Support. The Effective Utility Management Primer is a descriptive guide that provides a framework for assessing utility management practices according to the Ten Attributes. Areas of the PSW self-assessment process include consideration of topics that fall within several of the Ten Attributes areas. Completion of the PSW self-assessment process can be used as a tool to support and enhance a water utility's EUM assessment, strategy, and progress. Additional information and EUM resources may be obtained online at www.watereum.org.

Because considerable overlap exists between EUM and many aspects of the Administrative portion of the self-assessment process, this chapter includes a variety of references to EUM performance benchmarks. Utilities are encouraged to consider these benchmarks during completion of the Administrative assessment. Additional water and wastewater utility benchmarking information may be accessed in AWWA's Benchmarking Surveys (AWWA 2016), which are periodically conducted by AWWA. AWWA periodically publishes survey results so that utilities may compare performance with that reported by survey participants.

ADMINISTRATIVE POLICIES

Understanding

Often, the impact of administrative limitations is difficult to discern. The effect of administrative practices on performance or operations is not always as direct as it is with the design or operational factors assessed in previous chapters of this guide. For a utility to strive for optimized performance, a demonstrated commitment to excellence needs to be in place at all levels of the organization. This commitment must be based on an understanding of the importance of utility performance to the protection of public health. The distribution system is, in essence, the final barrier in the multi-barrier approach to protecting water quality and public health. Administrators must be willing to pursue actions aimed at improving distribution system performance and consistently demonstrate an understanding both of the health implications associated with operating a distribution system and of the utility's responsibility to be fiduciary stewards of the revenue collected from customers.

Administrators should provide direction on developing goals for maintaining high-quality water and emphasize to the operating staff the importance of achieving these goals. Goals should be understood and accepted by staff at all levels of the organization, and administration should empower its operating staff to achieve these goals. Utilities may consider adopting the PSW optimization goals for key distribution system parameters, and may also consider setting relevant internal performance goals for additional parameters. Note that any regulatory requirements that are more stringent than the PSW goals take precedence over the recommendations provided in this guide.

A significant challenge for administrators is to balance water quality, quantity, and cost to customers. Administrators should lead with an awareness of the utility's objective to achieve and consistently deliver high-quality water to its customers. Administrators should apply the Partnership optimization tools to improve working conditions and operational practices, support special studies, provide adequate numbers of qualified staff, embrace the Partnership goals, and maintain a rate structure that supports the utility's optimization activities. Optimization efforts may result in reduced operating costs, which can help to support and provide a basis for continuing optimization activities, although cost-savings should not be the sole objective of undertaking the optimization process.

Minor modifications identified in an action plan by utility staff during the self-assessment process, and the potential costs associated with these improvements, can often serve as a basis for assessing administrative factors that may be limiting optimized performance. For example, distribution system staff may have correctly identified needed minor modifications for facilities and presented these needs to the utility manager but had their requests declined. However, the self-assessment team must also solicit the other side of the story from the administrators to see if other factors contributed to the decision. There have been instances in which operators or field superintendents have convinced administrators to spend money to "correct" problems only to find that no improvement in distribution system performance resulted. Administrators also should be cognizant of the potential impact of utility policies on the optimization process.

Another area in which administrators can significantly, though indirectly, affect distribution system performance is through personnel motivation. A positive influence exists if administrators encourage personal and professional growth through support of training, both in-house and external opportunities; involvement in professional organizations, such as AWWA; tangible awards for initial certification or upgrading of certification levels; and encouraging similar professional growth activities.

It is also critical for utility management to be part of the optimization team. From the preceding discussion, it should be clear the integral role that utility administration plays in achieving distribution system optimization. For example, it is important for utility management to discourage complacency, especially when the utility relies on a high-quality finished water consistently entering the distribution system a high percentage of the time or a newer and well-maintained distribution system. Development of an environment that fosters a commitment to excellence can be the best defense against complacency and can help enable staff to be prepared to address rare and potentially serious events that have the potential to impact distribution system operation and performance. This requires the involvement of all staff members to create an empowered staff that can effectively respond to changing conditions, and development of an appropriate training program to ensure that operator skill is maintained when response to suboptimized distribution system conditions is required.

Self-assessing how the utility's administrative policies actually support the distribution system's performance goals is potentially very challenging. This self-assessment section should be conducted similarly to other sections of this guide, with involvement from the entire distribution system self-assessment team. The utility should openly embrace feedback and improvement opportunities presented by front-line staff regarding administrative policies that have the potential to improve the overall water quality delivered to customers.

Status

The status of administrative policies, as they relate to achieving a desired level of distribution system performance, may be assessed by reviewing the following self-assessment questions. The self-assessment team is not limited to these questions and may consider discussing additional topics related to administrative policies that may assist in identifying and addressing performance-limiting factors.

- Do the utility's overall goals and commitment, at the highest level of management, indicate a focus on supplying the highest quality of water possible to customers, and include the role of distribution system operators in achieving these goals?
  - ▲ The utility should have a documented and followed mission statement identifying water quality (i.e., public health) as the highest priority of the utility.
- Does the utility have a strategic planning process? Such a process should include:
  - ▲ Financial planning
  - ▲ Long-range facility planning
  - ▲ Capital replacement plans
  - ▲ Water quality master plans
- Does operating staff have authority to make required operation and maintenance decisions?
- Do management styles, organizational capabilities, or communication practices at any management level adversely affect performance?
  - ▲ Management objectives should encourage empowerment of staff to make decisions.
  - ▲ Organizational structure should support achieving water quality goals.
  - ▲ There should be clear alignment between distribution system goals and typical operations.
  - ▲ Utility staff should have regular meetings to discuss operating issues, water quality concerns, or other issues. These discussions

could occur at operator meetings that are regularly scheduled. This discussion may be more sustainable when a formal meeting is scheduled and a commitment is made from all stakeholders to attend and participate.

- Do administrators have a firsthand knowledge of distribution system needs, through site visits or discussions with utility staff?

  ▲ Administrators should be aware of water quality issues within the distribution system.

  ▲ Administrators should be accessible to utility staff.

- Does the utility have ongoing public information activities, including communications with customers on water quality issues and improvements?

  ▲ It is very important to foster a discussion with customers regarding water quality improvements before a problem occurs.

- Has the utility maintained awareness of existing and impending water quality regulations, and is this information used to develop long-term plans to ensure compliance?

- Does management support utility staff involvement in professional organizations?

  ▲ Staff should be encouraged to participate within the industry to learn new developments and be knowledgeable in current distribution system advancements. In addition, an optimized utility management team will encourage not only participation, but also leadership by staff at all levels within the utility, in professional organizations, such as AWWA, to give knowledge back to the industry, helping to better other utilities.

- Does management prioritize water quality over water quantity?

  ▲ Cost-effective water quality objectives should receive the highest priority in advance of water quantity initiatives, where practical. There can be a delicate balance between meeting the customer's water quantity needs and water quality expectations, particularly when additional water sales can generate the revenue needed to finance utility projects.

Action

If areas of the utility's administrative policies are considered to have a status of Partially Optimized or Not Optimized, and this parameter has been prioritized and selected for improvement, the following steps are recommended to be considered in the development of an action plan to reduce the identified limitations and improve performance in this area. This list is not comprehensive and staff is encouraged to develop utility-specific actions to address performance-limiting factors identified in this area.

- Develop mission and vision statements and a strategic plan.
- If, at any level in the utility, there is a lack of understanding of the need to supply the highest quality water possible or a lack of commitment to do so, develop an action plan to educate utility staff about the mission and vision of the organization. Part of this action plan should address specifically how the education will be communicated throughout the utility.
- The utility should make sure that there is a formal policy or process to ensure that frontline staff has the authority and capability to change operations to systematically respond to periods of poor water quality or operational excursions. This policy should also include procedures to follow when water quality deteriorates to the point that public notification may be required.
- The utility should support this policy with a training program that ensures that the empowered staff has the tools, compensation, knowledge, skills, and staffing levels necessary to respond to water quality changes in the distribution system.
- If there is a lack of understanding of existing and impending regulations, develop a list of regulatory requirements and assess their impact on the utility.
- The utility's management and operations staff perspectives should not differ significantly with regard to the needs of the distribution system. Procedures should be demonstrated to be in place that facilitate staff input into the budgeting process, both capital and operations.
- Communications procedures within the utility, at any level, must foster a positive impact on water quality. The type and frequency of communications should be detailed in the self-assessment report, including both internal and external communications

ACCEPTANCE OF OPTIMIZATION GOALS

Understanding

The Partnership's primary distribution system optimization goals, outlined in chapter 2, were the result of extensive research and scientific investigation to minimize the risk of waterborne disease. The utility may also choose to initially develop internal goals as a starting point and later adopt the Partnership goals as distribution system operations and personnel continue to progress toward true optimization. Regardless of how the goals are chosen, these goals should be reflected within operational guidelines, with demonstrated alignment between the stated goals and the manner in which the system is operated and managed. Furthermore, this commitment to the defined goals should be evident even when operations are challenging. If goals are abandoned during times of difficult operations, the utility needs to step back and evaluate whether goals are understood and embraced by all operations staff, as well as staff at all levels of the organization. If there is clear understanding of the optimization goals' importance as they relate to public health, this should help to enable moving the utility culture in that direction.

A basic tenet of the Partnership program is the acceptance of optimization goals. Acceptance means working toward, and eventually maintaining, the stringent goals set by the Partnership program during normal operations as well as during challenging or rapidly changing conditions. The acceptance of goals by utility staff needs to happen at all levels throughout the organization. Support from all levels in the utility, including the administration, needs to be secured for a utility to truly become optimized. This effort is important when there is time for optimization work and even more important when there is little time for it because of challenges in the distribution system. The self-assessment team needs to include a representative from administration to provide the overall management perspective to optimization and the self-assessment process. This integration is critical to demonstrate in the self-assessment completion report.

Status

The status of universal acceptance of the optimization goals within a utility, as related to achieving a desired level of distribution system performance, may be assessed by reviewing the following self-assessment questions. The self-assessment team is not limited to these questions and may consider discussing additional topics related to the acceptance of optimization goals that may assist in identifying and addressing performance-limiting factors.

- Does operational guidance reflect the optimization goals adopted by the utility?

- Does operations staff maintain a focus on goals even during challenging water quality and/or operational events that require changes to specific operating parameters?

  ▲ One test of goal acceptance is the actions taken by utility staff during events that may pose a challenge to water quality and operations.

  ▲ An example of this is when distribution system staff manipulates tower levels in the extremities of the distribution system to maintain water quality, even during periods of high demand. The focus on water quality, adhered to even in the face of multiple system challenges, demonstrates staff acceptance of and commitment to the utility's optimization goals.

- Does administrative staff embrace and pursue goals "beyond the regulations"?

  ▲ The self-assessment team should document examples of the pursuit of optimization goals "beyond regulations." There should be clear indication that all system personnel are actively striving to meet interim goals now, and ultimately the Partnership's optimization goals in the future.

- Does the organization's mission statement embody the Partnership philosophies?

  ▲ The utility should be able to relate the meaning of its mission statement to specific Partnership optimization principles.

- Is distribution staff empowered to make operating changes to address potential water quality issues or are proposed changes required to be approved by management?

- Can appropriate changes be made during the evenings, weekends, and holidays?

Action

If the acceptance of optimization goals is considered to have a status of Partially Optimized or Not Optimized, and this parameter has been prioritized and selected for improvement, the following steps are recommended to be considered in the development of an action plan to reduce the identified limitations and improve performance in this area. This list is not comprehensive, and staff is encouraged to develop utility-specific actions to address performance-limiting factors identified in this area.

- After careful evaluation of the questions presented here, determine the best approach to ensure that the optimization goals are clearly in alignment with utility policy and procedures, especially during challenging circumstances.

- If there is a disparity between the goals and how the system is operated, develop an action plan to meet with the distribution system operations team to reach a resolution. This may require some challenging conversations with management, SCADA system modifications, or rewriting of operational guidance. The effort will be well worth it, as the operational consistency and operator empowerment will far outweigh the effort required.

- The team should review SOPs and operational guidance, deliberately incorporating the distribution system goals into these documents.
- Consider developing operational "safety nets" within the control system that are aligned with distribution system goals.

INVOLVEMENT OF ALL PARTIES IN THE PARTNERSHIP PROCESS

Understanding

As described in previous chapters, the self-assessment team optimally should include personnel from all disciplines and levels of the organization, including management. Involvement of all parties in the Partnership process develops and nurtures personal and organizational ownership of, and commitment to, common goals that promote the best opportunity for success. Strong and committed leadership is an important key to establishing, cultivating, and directing this team effort to ensure all parties remain committed to and focused on optimization goals. Using the collective knowledge, experience, and wisdom of all parties is simply the best and most efficient way to achieve excellence. The ability to identify and successfully address and correct issues and provide guidance that enables all personnel to work to achieve the same objective will facilitate this process. The best opportunity for success requires administrators who involve staff at all levels to develop and set goals for the organization as well as promote an environment where all parties are committed to excellence and expected to participate in the optimization process.

Status

The status of the involvement of all parties in the Partnership process, as related to achieving a desired level of distribution system performance, may be assessed by reviewing the following self-assessment questions. The self-assessment team is not limited to these questions and may consider discussing additional topics related to the involvement of all parties in the Partnership process that may assist in identifying and addressing performance-limiting factors.

- Has the organization developed a vision and/or mission statement that incorporates the PSW process?
  - ▲ The vision and/or mission statement should clearly highlight the core vision and mission of the organization. The development and periodic review of the vision/mission statement will ideally provide an opportunity for all employees to reflect on what is most important, offer input to define or refine the vision and/or mission, and begin or reinforce the process of involvement of all parties in the PSW process.

- Is there strong leadership throughout the organization that supports the PSW culture?
  - ▲ Leadership is a process whereby an individual (or individuals) influence others to achieve a common goal. Ideally, leadership exists at all levels of an organization. The ability to establish and cultivate strong leadership in the organization will help to ensure that near- and long-term optimization goals are developed and met.
  - ▲ Effective leadership communication helps employees understand the utility's overall business strategy, how they contribute to achieving optimization goals, and by sharing information with employees on how the organization as a whole and the employee's own division are working toward optimization.
- Have all personnel been briefed on the purpose of the Partnership Program and do they have a good understanding of the overall intent to optimize distribution system operation, as well as the benefits of being a member of the Partnership?
  - ▲ Employees should be aware of the data collection required, data interpretation, and the use of data as a tool to measure progress toward optimization. One of the keys to obtaining buy-in is to quantify performance improvements through the data interpretation process.
  - ▲ Employees should be encouraged and willing to seek out and identify performance-limiting factors, recommend specific changes, and become actively involved in the development and implementation of action plans that mitigate or eliminate them. Remember that the level of involvement of the team is directly related to employees feeling they are a valuable part of the process and that their views are heard.
  - ▲ Employees should be trained to be vigilant and remain tenacious in their awareness of actions and situations that may adversely impact distribution system goals and compromise optimization. These scenarios should be reviewed and proper responses documented.
- Does each employee feel that his or her role is well defined and do employees understand their roles in the optimization process?
- Do administrators set goals and provide an environment that expects and encourages involvement in optimization at all levels?

Action

If the involvement of all parties in the Partnership process is considered to have a status of Partially Optimized or Not Optimized, and this parameter has been prioritized and selected for improvement, the following steps are recommended to be considered in the development of an action plan to reduce the identified limitations and improve performance in this area. This list is not comprehensive, and staff is encouraged to develop utility-specific actions to address performance-limiting factors identified in this area.

- Evaluate the specific ways the organization promotes, supports, and sustains a culture of excellence, and ensures involvement of all parties in the Partnership process. Revise documentation, if necessary.

- Develop or revise the utility's vision and/or mission statements, involving all parties in the process.

- If there are problems with leadership/administration, identify the problem(s) and develop action plans to address them.

- If there is a lack of understanding of the Partnership process and goals, develop action plans to ensure all parties obtain this knowledge through proper training and practice.

- Be sure to include members from all levels of the organization on the self-assessment team. This will help to ensure representative responses to all of the self-assessment questions.

- Develop methods and training to cultivate institutional tenacity and awareness of complacency issues.

DOCUMENTATION/DEMONSTRATION OF ADDRESSING COMPLACENCY

Understanding

Complacency will lead to nonoptimized distribution system operation. Complacency is defined as the feeling of security while being unaware of potential dangers or defects that pose a potential risk to water quality and therefore to public health. The unrelenting struggle to address complacency involves skills, policies, and procedures intended to enhance problem-solving and address rare and potentially serious events that may lead to extreme and prolonged water quality degradation. The primary objective of documenting and addressing complacency is to demonstrate (1) the utility's capability to address rare and potentially serious events, (2) enhanced problem-solving by front-line personnel so that taking action is not limited to managers or other nondirect operations personnel that do not have proximity to quickly move to change processes and procedures when conditions call for change, and (3) an internal ongoing process whereby developing situations are used to plan for required operational changes and past experiences lead to beneficial gains in anticomplacency measures.

One of the situations in which complacency should be carefully considered, and steps taken to avoid it, is when a utility is facing a potential change in source water, which can occur for a variety of reasons. As described previously in chapter 3, source water changes have the potential to significantly impact treated and distributed water quality. If a source water change is to be considered by a utility, its potential impact should be carefully evaluated through desktop, bench, and pilot-scale testing, with attention paid to simultaneous compliance issues. It is also recommended to discuss potential source water changes with the utility's local regulatory agency, prior to their implementation. Refer to chapter 3 for a detailed discussion of the impact of selected water quality parameters on overall distribution system water quality.

Complacency can also act to dull the response of a utility to changes in water quality resulting from changing conditions, such as drought or variations in customer water demand. Drought conditions may require pumping from a water source that exhibits differences in water quality from the system's primary raw water source. The new source water may meet all water quality goals at the well or treatment plant; however, a change in pH or alkalinity can have a significant effect on disinfectant residual or the potential for corrosion in the distribution system. Water chemistry changes, over time, can be subtle and need to be recognized and tested to determine the impact to distribution system water quality.

Similarly, increases or decreases in water demand can affect how water behaves in the distribution system piping. Water age can vary because of changing flow conditions caused by a reduction in water demand; for example, decreased water usage can result in increased water age in the system. This can lead to nonoptimized, low disinfectant residual concentrations in the system. These changes can be subtle and, when combined with complacency within the utility, can cause a great reduction in customer water quality and protection.

Table 6-1 shows best practices that optimized utilities implement to avoid complacency. A self-assessment examines all aspects of distribution system operation, including problem-solving and decision-making. A variety of distribution system practices may be implemented to maintain a proactive approach and avoid complacency. Thesecan include establishing sustainable procedures to optimize production and meet system demand under challenging conditions without sacrificing water quality.

In addition, utilities should strive to identify rare and potentially serious events, specific to the unique aspects of the utility's source, service area, and/or treatment technologies, and to develop policies and procedures for the distribution system to mitigate events that may lead to nonoptimized performance.

Status

The status regarding complacency, as related to achieving a desired level of distribution system performance, may be assessed by reviewing the following self-assessment questions. Complacency may also be assessed by reviewing the thoroughness and timeliness of submitted standard operating procedures (SOPs) that help to improve water quality and combat complacency through routine and emergency activities. The self-assessment team should also review the best practices listed in Table 6-1 to determine whether these factors are implemented at the utility to guard against complacency. The self-assessment team is not limited to these questions and may consider discussing additional topics related to complacency that may assist in identifying and addressing performance-limiting factors. The self-assessment completion report should include a description of actions and steps taken to prevent complacency.

- Does the utility employ the best practices listed in Table 6-1 to combat complacency?

- During the past 10 years, how has staff responded to a major episode of diminished water quality in the distribution system?

- If complacency was determined to be the cause of a water quality episode, what actions have been taken to prevent it from occurring in the future? What process control procedures and special studies were completed related to the identified cause of complacency?

- How would staff respond to deteriorating water quality during unusual changes in water quality in the distribution system?

- How would staff address a situation in which all of the usual operational control procedures do not maintain optimized performance?

- Does the utility have a policy/procedure to implement when considering significant changes to the type of source water or water quality entering the distribution system that considers the impact of water quality changes on overall distribution system water quality?
  - ▲ Parameters to consider include alkalinity/pH, pipe materials, disinfectant residual, etc.
  - ▲ Refer to the Internal Corrosion Control section of chapter 3 for a more detailed discussion of how these parameters can affect distribution system corrosivity and overall water quality.

Action

If complacency is considered to have a status of Partially Optimized or Not Optimized, and this performance improvement variable has been prioritized and selected for improvement, the following steps are recommended to be considered in the development of an action plan to reduce the identified limitations and improve performance in this area. This list is not comprehensive, and staff is encouraged to develop utility-specific actions to address performance-limiting factors identified in this area.

- Develop operational guidance to respond to episodes of poor water quality.
- Consider developing process control procedures and performing special studies to identify and address causes of complacency.

- Develop SOPs for staff response to deteriorating water quality and unusual water quality events.

- Ensure there is a strategy for handling situations in which all of the usual operational procedures do not maintain optimized performance, including specific roles of personnel.

Table 6-1. Partnership for Safe Water—best practices for avoiding complacency

Complacency definition: Inadequate capability (i.e., skills, policies, procedures) exists to maintain optimized performance during nonroutine events and/or rare and potentially serious events.
Practices of optimized utilities to help prevent the existence of complacency:
• Maintaining utility awareness and commitment for public health responsibilities and the need to consistently achieve performance goals.
• Developing formal policies and procedures to implement source selection when this capability exists.
• Having a means for identifying and considering algal blooms and their impact on distribution system performance and operation.
• Demonstrating an understanding of complacency (i.e., optimized systems do not take for granted their performance status nor deny the future potential for challenging situations and/or potential inability to successfully handle challenging situations).
• Maintaining a priority on water quality over short-term water demand needs.
• Understanding of the utility's potential vulnerabilities and how they would be addressed (i.e., an outbreak has never happened here, but could it occur in the future and how would it be addressed?).
• Embracing change and developing a formal operational control program that anticipates future changes.
• Making data-driven operational control decisions, based on established and documented processes, not made solely based on the informal, nondocumented "experience of the staff." This can relate to the active management of water storage facilities. A data-driven process can help in setting up a control scheme that best suits water quality needs.
• Policies and procedures exist to address all potential events that could affect water quality. This may include the use of interconnections with other systems, where appropriate.
• Selecting distribution system operational strategies that are appropriate to address changes in water quality, either those that occur frequently or rarely, identifying and controlling any potential risks to the best of the utility's ability.
• Good water quality is primarily the result of thorough planning, documented procedures, and anticipation of all potential rare and potentially serious events that may impact water quality.
• Special studies/characterization technologies (DNA testing of biofilms, biome characterization, special residual studies, tracer testing, etc.) may be used for problem-solving to address issues that challenge, or could challenge, the distribution system.
• Serving consecutive systems does not mean a utility's reaction time can be shortened or reduced because the problem has "moved to the next community."

TRAINING

Understanding

A robust training program is a critical component for any organization to ensure that its goals are consistently achieved and to provide an avenue to enhance the skill set of the employees. A strong message is relayed throughout the utility about its values and mission when an emphasis is placed on proper and consistent training. By investing in employee training, the utility will be rewarded with a more highly skilled workforce, improved employee safety, and greater morale brought about by the attention given to the employees' professional growth.

All employees, regardless of experience, should receive regular training that covers a variety of areas such as safety, regulatory issues, operational procedures, and emergency response. The overall program should include orientation for the new employee, refresher training for veteran employees, and procedures to evaluate the effectiveness of the training activities. Emergency response training should be included: an employee needs to address emergency situations with calmness and self-assurance that are the result of constant review and reinforcement of the appropriate actions.

Training should not be approached as a "one and done" but rather a constant mindset of continual improvement. Not all learning comes from a formal classroom setting; however, the training program should be intentional and well documented. Attention should also be given to the quality of training: substandard efforts that are often present when the attitude is to "check a box" and mark a task complete can be a waste of time and resources.

The time a utility spends in training will reap great dividends in producing high-quality drinking water, maintaining a safe work environment, and promoting employee confidence. A comprehensive training program should be emphasized in the organization to adequately educate every employee. The utility should embrace and empower employees to become certified operators in the distribution system within their local certification program. Some areas may have varying levels of operator certification. The utility should encourage employees to achieve a certification level higher than their current job duties require.

Status

The status of training, as related to achieving a desired level of distribution system performance, may be assessed by reviewing the following self-assessment questions. The self-assessment team is not limited to these questions and may consider discussing additional topics related to training that may assist in identifying and addressing performance-limiting factors.

- Does training/documentation exist for new employees?

  ▲ The employee should receive training on the distribution system's standard operating procedures.

  ▲ The employee should demonstrate competency of required job skills.

  ▲ Documentation should exist to support training activities and competencies.

  ▲ A training schedule should be documented, covering the introduction period for new employees. This period will vary for each utility, but will likely be at least 1 year. The onboarding training should provide necessary training to bring the new employee up to speed with veteran employees.

- Does training/documentation exist for all employees?

  ▲ Employee's job performance should be reviewed to confirm ongoing understanding and skills to achieve the utility's operational and safety goals. This review also serves to verify that the employee is adhering to accepted procedures and that shortcuts have not been developed that could jeopardize distribution system operations.

  ▲ Emergency response to such items as contamination events, atypical water quality conditions, equipment failure, and other vulnerabilities should be reviewed and practiced.

  ▲ Performance evaluation standards should be used to verify water quality testing accuracy and reliability.

  ▲ A program to track and address continuing education requirements for operator licenses should be established.

  ▲ A formal budget line item should be established to ensure adequate funding is allocated for employee training needs.

  ▲ An annual schedule of training activities should be documented for the utility.

  ▲ Annual, or more frequent, SOP review and training should be specifically emphasized.

  ▲ A documented training program will include a description of tailgate or morning training sessions based on activities planned for the day.

- Does the utility ensure that employees receive appropriate safety training?

  ▲ Safety training should include topics required by law as well as those that are deemed as safe operating practices for the individual system. It may be beneficial to have the employees participate in a job safety analysis for their specific job tasks.

- Is the distribution system staff knowledgeable about drinking water regulations and emerging technologies?
  - ▲ Drinking water regulations and water industry publications should be readily available to the employees.
  - ▲ Informal review of drinking water topics, at tailgate meetings or in the course of the workday, is recommended.
  - ▲ Employees should regularly attend training seminars and continuing education courses to update knowledge of drinking water issues.
- Does a culture of learning and continuous improvement exist throughout all levels of staffing?
  - ▲ The utility should encourage and develop an atmosphere of learning that enables all employees to achieve their highest potential.

Action

If training is considered to have a status of Partially Optimized or Not Optimized, and this parameter has been prioritized and selected for improvement, the following steps are recommended to be considered in the development of an action plan to reduce the identified limitations and improve performance in this area. This list is not comprehensive, and staff is encouraged to develop utility-specific actions to address performance-limiting factors identified in this area.

- Development of an orientation manual for new employees.
- Use of evaluation forms to confirm and document training activities.
- Confirmation of water quality testing competencies with performance evaluation standards.
- Routine review of procedures with each employee to ensure consistency of tasks with all staff.
- Development of tabletop scenarios and exercises to practice and evaluate emergency response.
- Confirmation that all required safety training is being conducted. This may include, but is not limited to, Occupational Safety and Health Administration regulations or similar safety regulations as well as the utility's safety or risk management code.
- Provide easy access to educational materials, such as *Journal AWWA*, *Opflow*, and AWWA manuals and videos.
- Participation in training classes and seminars offered by outside agencies.

STAFFING

Understanding

Utility staffing must be sufficient to maintain distribution system performance goals on a continuous basis. Proper staffing is critical with respect to the levels of distribution system monitoring necessary to ensure that performance goals are achieved on a continuous basis. For an optimized distribution system, with respect to staffing, staffing levels should optimally allow 24-hour-per-day, 365-day-per-year coverage of the distribution system performance elements, such as pressure and disinfectant residual, which can be achieved through SCADA and remote monitoring technologies. In addition, a utility should have a documented emergency response time goal, which could be tiered based on the severity of the performance deviation. Adequate staffing should take into account allowances for leave, as well as anticipated retirements.

Some utilities use instrumentation with alarms as a substitute for 24-hour physical coverage. There are many beneficial automation tools that exist for monitoring and controlling distribution systems, however it is important not to completely replace human judgment and operational expertise with the control system. A utility should document how automation is used to supplement the distribution system operator's expertise. Refer to ANSI/AWWA Standard G200—Distribution System Operations and Management for more detailed information regarding the use of controls to support unstaffed utility operation.

A variety of utility benchmarking performance indicators may be used to quantify utility staffing. These include common indicators referenced in AWWA Benchmarking Surveys, as well as those referenced in EUM resources. Although an evaluation of specific staffing performance benchmarks is beyond the scope of this self-assessment guidance, utilities are encouraged to consider assessing the benefits of whether the evaluation of staffing benchmarks may yield any additional benefits for the utility.

Status

The status of staffing, as related to achieving a desired level of distribution system performance, may be assessed by reviewing the following self-assessment questions. The self-assessment team is not limited to these questions and may consider discussing additional topics related to staffing that may assist in identifying and addressing performance-limiting factors.

- Are crews of sufficient size to accomplish necessary maintenance and repair activities in a safe and timely manner?
- Are response time goals (state goals) consistently met?

- Is the number of distribution system staff adequate to maintain optimized performance?

  ▲ Consider evaluating staffing benchmarks to further assess performance, such as:

    ▪ Customer accounts per employee: Number of accounts/number of FTEs (FTE = 2080 hours per year of employee time equivalent)

    ▪ MGD of water delivered/processed per employee: Average MGD delivered/FTEs per year.

  ▲ Distribution system staff adequacy may also be assessed by considering the following:

    ▪ Ensuring adequate coverage for routine operational tasks as well as emergency activities.

    ▪ Utilities with staff that has accumulated significant vacation benefits may require more operators.

    ▪ Percentage of staff and leadership positions with defined competencies.

- Do all distribution system operators possess the appropriate level of certification/licenses to operate the distribution system?

  ▲ Temporary or part-time workers, used to fill in schedules, must have sufficient skills and experiences.

  ▲ Temporary staff may need to be shadowed by a full-time staff member to ensure consistency with operational goals and procedures.

  ▲ For large and medium utilities, a water quality professional position and/or a separate water quality department indicate a commitment to a proactive, distribution water quality program.

- Does the utility have the instrumentation to continuously monitor all critical parameters with alarms? This of particular importance if the system is operated without staff present (note: although operating without staff present may not be a desired option, having remote, on-call staff that can be notified of and appropriately respond to distribution system events may be applicable).

  ▲ Review emergency response staffing for weekends, holidays, and vacations for periods of unstaffed operation.

  ▲ If the distribution system is operated without continuous staffing, alarms are recommended, at a minimum, for disinfectant residual and pressure readings outside of minimum or maximum setpoints. Similar alarms are recommended on all processes and/or parameters critical to distribution system operation. These alarms may be addressed by the utility's treatment plant or by call out to emergency on-call staff, within an appropriate time frame to avoid distribution system conditions approaching unacceptable levels.

▲ If online instrumentation and alarms are depended on to protect distribution system water quality, a quality control program/preventive maintenance program, conducted by skilled personnel, must be in place. Additional information regarding instrument maintenance is provided in chapter 5.

- Does the current staffing level have a detrimental effect on distribution system operation, maintenance, or sampling procedures?

   ▲ A sufficient number of people needs to be available to perform distribution system operations activities, such as maintaining disinfectant residual and pressure standards, monitoring pumping activities, and managing customer expectations.

   ▲ Determine if use of excessive overtime and an inability to take vacations exists. Review overtime records to determine if excessive overtime is being accumulated.

   ▲ Determine how emergency measures will be immediately implemented should an operator become incapacitated.

- Does a high staff turnover rate, high degree of absenteeism, or large number of grievances indicate other underlying staffing problems?

   ▲ Consider evaluating staffing benchmarks to further assess performance, such as:

      ■ Employee Turnover Rate (%): 100 * (number of employee departures/total number of authorized positions per year)

      ■ Retirement Turnover (%): 100 * (number of retirement departures/total number of authorized positions per year). This measure examines the potential vulnerability of the distribution system due to loss/retention of institutional knowledge

      ■ Experience Turnover (%): 100 * (number of years of experience represented by all departures/total years of experience with the organization). Calculate and evaluate this measure at the beginning of each year.

- Does the improper distribution of adequate staffing result in poor distribution system performance or response to performance deviations?

   ▲ The improper distribution of adequate staffing can prevent system adjustments from being made or cause them to be made at inappropriate times, resulting in less than optimized performance levels.

- Does a low pay scale or benefit package discourage more highly qualified personnel from applying for utility positions or cause personnel to leave once they are trained?

  - Pay scales and benefits should be commensurate with other utilities in the area to attract and maintain a professional staff.

  - Consider evaluating staffing benchmarks to further assess performance, such as:

    - Employee salary competitiveness relative to the market rate: Average percentile rank of employee salaries compared with salaries in surrounding service areas, as determined by a market rate comparison.

Action

If staffing levels are considered to have a status of Partially Optimized or Not Optimized, and this parameter has been prioritized and selected for improvement, the following steps are recommended to be considered in the development of an action plan to reduce the identified limitations and improve performance in this area. This list is not comprehensive, and staff is encouraged to develop utility-specific actions to address performance-limiting factors identified in this area.

- If the number of staff available to operate and maintain the system and conduct necessary sampling operations is insufficient and is the cause of poor distribution system performance, develop an implementation plan to add staff or to address the limitation through other means, such as shared personnel with other utilities.

- If the number of staff is adequate but that staff is distributed improperly to obtain optimized distribution system performance, develop an implementation plan to redistribute the workforce to support consistent achievement of water quality goals.

- If distribution system staff does not possess the required or anticipated levels of certifications, develop a training plan that allows operators to train and test for their required level of certification.

- If instrumentation and controls are determined to be inadequate to support the desired staffing levels, consider improving the systems monitoring instrumentation and control system to provide the information required to support staffing levels. Refer to chapter 5 for additional information.

FUNDING

Understanding

Financial resources are required to pursue high-quality water as a utility goal. Adequate rate structures provide the funding for obtaining professional operating, maintenance, and laboratory staff, training of these personnel, and maintaining equipment and existing facilities necessary to provide high-quality water on a continuous basis. When reviewing financial information, the impact of bonded indebtedness on the utility should be determined, as should whether the rate structure creates sufficient revenue to adequately support the utility given the indebtedness.

An administrator should strive to create and maintain a self-supporting utility; however, sometimes managers of small utilities create enough debt to enable the utility to be eligible for government grants. This can be especially damaging to the long-term stability of the utility and distribution system performance because it allows few options for financing improvements that may be necessary to meet current and/or future water system regulations.

Utility funding is often maintained at levels too low to optimize performance because of the desire to avoid rate increases. Effective utility managers have the ability to overcome perceived resource barriers by properly prioritizing resource allocations so that they can be proactive in developing adequate rate structures to support goals of high water quality.

A variety of utility benchmarking performance indicators may be used to quantify utility funding. These include common indicators referenced in AWWA Benchmarking Surveys, as well as those referenced in EUM resources. Although an evaluation of specific funding performance benchmarks is beyond the scope of this self-assessment guidance, utilities are encouraged to consider assessing the benefits of whether the evaluation of specific financial benchmarks may yield any additional benefits for the utility.

Status

The status of funding, as related to achieving a desired level of distribution system performance, may be assessed by reviewing the following self-assessment questions. The self-assessment team is not limited to these questions and may consider discussing additional topics related to funding that may assist in identifying and addressing performance-limiting factors.

- Does the utility have the financial health to be self-sufficient?
- Do revenues cover expenses, and is there sufficient funding to cover unexpected expenditures?

> ▲ Consider evaluating funding benchmarks to further assess short and long term performance, such as:
>
> - Revenue to Expenditure ratio: Total revenue/total expenditures
> - Operations and Maintenance (O&M) Expenditures (%): 100 * (O&M expenditures/total operating budget)
> - Capital Expenditures (%): 100 * (capital expenditures/total capital budget)
> - Debt Ratio: Total liabilities/total assets
> - Current level of operating reserves as a percentage of goal
>
> ▲ In assessing the ability of the current rate structure to adequately support distribution system activities, consider the following:
>
> - Adequate staffing to meet water quality goals
> - Capital improvement and replacement plans
> - Funding to allow for adequate training for operators, management, and laboratory personnel
> - Adequate staffing and equipment for a preventive maintenance program
> - Funding of capital reserve fund in the annual operating budget

- Review all major expenditures over the past 5 years and planned future expenditures. Do past and planned expenditures support the utility's water quality goals?

- Does the utility's bond indebtedness limit funds available for other needed items?

  > ▲ Interest payments should be less than 25% of the utility's budget.
  >
  > ▲ Debt should be used for long life improvement and replacement projects that occur infrequently, such as water tower replacement/ installation.
  >
  > ▲ Debt should not be used for ongoing projects, such as water main replacement programs.

- Is a declining population or water use trends expected to reduce water usage and revenues? Review future projections so that these situations can be accounted for in future planning.

- Does the utility calculate values for performance indicators? Does the utility perform comparisons to external financial benchmarks?

  > ▲ Although the physical and financial composition of every utility is unique, comparison to external benchmarks or internal benchmarks

over time can provide indicators about the financial health of the utility.

- Are utility rates sufficient to recover operations and maintenance expenses (minus depreciation expense) as well as provide for a healthy capital investment/reinvestment including debt service?

- Does the system have reserves to cover 6 months of O&M expenses?

Action

If funding is considered to have a status of Partially Optimized or Not Optimized, and this parameter has been prioritized and selected for improvement. The following steps are recommended to be considered in the development of an action plan to reduce the identified limitations and improve performance in this area. This list is not comprehensive, and staff is encouraged to develop utility-specific actions to address performance-limiting factors identified in this area.

- If there are revenue shortfalls, develop an action plan to develop a rate structure that is adequate to support all of the long-term needs of the utility. In this analysis and planning process, also include sufficient funds for unexpected expenditures or capital reserve funds.

- If there is a history of expenditures that do not support the utility's water quality goals, assess the decision making and prioritization process that contributed to these decisions.

- If interest payments caused by an excessive amount of bonded indebtedness negatively affect the availability of funds for key utility operations, develop an action plan to reduce the amount of long-term bonded indebtedness. Also, assess the use of bonds to finance anticipated future expenditures.

Performance-Limiting Factors Summary

Table 6-2 summarizes the administrative factors that may be limiting performance of the distribution system and utility overall. Check whether the factors in Table 6-2 are Optimized and Documented, Partially Optimized, or Not Optimized. Those factors checked as Partially Optimized or Not Optimized will be prioritized in chapter 7, Identification and Prioritization of Performance-Limiting Factors, so that an action plan may be created for optimization of parameters selected for improvement.

Table 6-2. Administrative performance-limiting factors summary

Self-assessment category	Questions for gauging optimization status	Optimized and documented	Partially optimized	Not optimized	Comments
Administrative Policies	Do the utility's overall goals and commitment, at the highest level of management, indicate a focus on supplying the highest quality of water possible to customers, and include the role of distribution system operators in achieving these goals?				
	Does the utility have a strategic planning process?				
	Does operating staff have authority to make required operation and maintenance decisions?				
	Do management styles, organizational capabilities, or communication practices at any management level adversely affect performance?				
	Do administrators have a firsthand knowledge of distribution system needs, through site visits or discussions with utility staff?				
	Does the utility have ongoing public information activities, including communications with customers on water quality issues and improvements?				
	Has the utility maintained awareness of existing and impending water quality regulations, and is this information used to develop long-term plans to ensure compliance?				
	Does management support utility staff involvement in professional organizations?				
	Does management prioritize water quality over water quantity?				

Self-assessment category	Questions for gauging optimization status	Optimization status			
		Optimized and documented	Partially optimized	Not optimized	Comments
Acceptance of Optimization Goals	Does operational guidance reflect the optimization goals adopted by the utility?				
	Does operations staff maintain a focus on goals even during challenging water quality and/or operational events that require changes to specific operating parameters?				
	Does administrative staff embrace and pursue goals "beyond the regulations"?				
	Does the organization's mission statement embody the Partnership philosophies?				
	Is distribution staff empowered to make operating changes to address potential water quality issues or are proposed changes required to be approved by management?				
	Can appropriate changes be made during the evenings, weekends, and holidays?				
Involvement of All Parties in the Partnership Process	Has the organization developed a vision and/or mission statement that incorporates the PSW process?				
	Is there strong leadership throughout the organization that supports the PSW culture?				
	Have all personnel been briefed on the purpose of the PSW Program and do they have a good understanding of the overall intent to optimize distribution system operation, as well as the benefits of being a member of the PSW?				

Self-assessment category	Questions for gauging optimization status	Optimization status			Comments
		Optimized and documented	Partially optimized	Not optimized	
Involvement of All Parties in the Partnership Process (*continued*)	Does each employee feel that his or her role is well defined and do employees understand their roles in the optimization process?				
	Do administrators set goals and provide an environment that expects and encourages involvement in optimization at all levels?				
Documentation/Demonstration of Addressing Complacency	Does the utility employ the best practices listed in Table 6-1 to combat complacency?				
	During the past 10 years, how has staff responded to a major episode of diminished water quality in the distribution system?				Provide description.
	If complacency was determined to be the cause of a water quality episode, what actions have been taken to prevent it from occurring in the future? What process control procedures and special studies were completed related to the identified cause of complacency?				Provide description.
	How would staff respond to deteriorating water quality during unusual changes in water quality in the distribution system?				Provide description.
	How would staff address a situation in which all of the usual operational control procedures do not maintain optimized performance?				Provide description.
	Does the utility have a policy/procedure to implement when considering significant changes to the type of source water or water quality entering the distribution system, that considers the impact of water quality changes on overall distribution system water quality?				

Self-assessment category	Questions for gauging optimization status	Optimization status			Comments
		Optimized and documented	Partially optimized	Not optimized	
Training	Does training/documentation exist for new employees?				
	Does training/documentation exist for all employees?				
	Does the utility ensure that employees receive appropriate safety training?				
	Is the distribution system staff knowledgeable about drinking water regulations and emerging technologies?				
	Does a culture of learning and continuous improvement exist throughout all levels of staffing?				
Staffing	Are crews of sufficient size to accomplish necessary maintenance and repair activities in a safe and timely manner?				
	Are response time goals (state goals) consistently met?				
	Is the number of distribution system staff adequate to maintain optimized performance?				
	Do all distribution system operators possess the appropriate level of certification/licenses to operate the distribution system?				
	Does the utility have the instrumentation to continuously monitor all critical parameters with alarms?				
	Does the current staffing level have a detrimental effect on distribution system operation, maintenance, or sampling procedures?				
	Does a high staff turnover rate, high degree of absenteeism, or large number of grievances indicate other underlying staffing problems?				

Self-assessment category	Questions for gauging optimization status	Optimization status			Comments
		Optimized and documented	Partially optimized	Not optimized	
Staffing (*continued*)	Does the improper distribution of adequate staffing result in poor distribution system performance or response to performance deviations?				
	Does a low pay scale or benefit package discourage more highly qualified personnel from applying for utility positions or cause personnel to leave once they are trained?				
Funding	Does the utility have the financial health to be self-sufficient?				
	Do revenues cover expenses, and is there sufficient funding to cover unexpected expenditures?				
	Review all major expenditures over the past 5 years and planned future expenditures. Do past and planned expenditures support the utility's water quality goals?				
	Does the utility's bond indebtedness limit funds available for other needed items?				
	Is a declining population or water use trends expected to reduce water usage and revenues?				
	Does the utility calculate values for performance indicators? Does the utility perform comparisons to external financial benchmarks?				
	Are utility rates sufficient to recover operations and maintenance expenses (minus depreciation expense) as well as provide for a healthy capital investment/reinvestment including debt service?				
	Does the system have reserves to cover 6 months of operations and maintenance expenses?				

Table 6-3. Administration and utility management references

Reference	Description
AWWA, 2015. G200-15, *Distribution System Operation and Management.* Denver, Colo.: AWWA.	A utility management Standard referencing best practices for distribution system operation and management.
AWWA, 2016e. *Benchmarking Performance Indicators for Water and Wastewater Utilities: 2016 Edition.* Denver, Colo.: AWWA. ISBN 9781625761972.	AWWA's Benchmarking Surveys provide an annual summary of key utility performance indicators for drinking water, wastewater, and combined utilities.
AWWA, 2016f. *2016 AWWA Compensation Survey: Large Water and Wastewater Utilities.* Denver, Colo.: AWWA. ISBN 9781625761552.	AWWA's annual compensation surveys include analyses of salaries, salary ranges, and compensation practices for water and combined water services utilities across North America.
AWWA, 2016g. *2016 AWWA Compensation Survey: Medium-Sized Water & Wastewater Utilities.* Denver, Colo.: AWWA, Denver. ISBN 9781625761569.	AWWA's annual compensation surveys include analyses of salaries, salary ranges, and compensation practices for water and combined water services utilities across North America.
AWWA, 2016h. *2016 AWWA Compensation Survey: Rural Water & Wastewater Utilities.* Denver, Colo.: AWWA. ISBN 9781625761576.	AWWA's annual compensation surveys include analyses of salaries, salary ranges, and compensation practices for water and combined water services utilities across North America.
AWWA, 2016i. *2016 Water and Wastewater Rate Survey - Book.* Denver, Colo.: AWWA. ISBN 9781625762078.	AWWA's annual water and wastewater rate surveys provide a comprehensive report of water and wastewater utilities rates in the United States. Reports available in hardcopy or electronic format.
Effective Utility Management: A Primer for Water and Wastewater Utilities, 2017. USEPA. 2017. Effective Utility Management: A Primer for Water and Wastewater Utilities. http://www.nacwa.org/docs/default-source/resources---public/eum-primer-final-1-24-17.pdf	Provides information about best practices for Effective Utility Management, along with procedures to assess current utility management practices and develop and implement improvement plants.

Note: additional supporting information may be accessed online at Work for Water (www.workforwater.com) and the Value of Water Coalition (www.thevalueofwater.org).

CHAPTER 7

CONTINUOUS IMPROVEMENT PLAN

Introduction

This chapter describes the process of identifying and prioritizing performance-limiting factors so that the utility may develop an optimization action plan. All utilities completing the self-assessment process should complete this chapter of the guide to prioritize factors limiting optimized performance and develop an action plan to help enable and realize improvement for the highest priority performance-limiting factors.

Completion of the self-assessment process to this point results in systems gathering information on distribution system performance and factors that could be limiting performance in the areas of design, operations, and administration. The next step in the self-assessment process is the evaluation of the information that has been collected and the identification and prioritization of the primary reasons for less-than-optimized performance. This step is critical in defining the focus of activities that must be implemented to move the system toward optimized performance. For this reason, the assessment team must continue to include individuals from all departments with a vested interest in realizing distribution system improvements, including: operators, supervisors, managers, water quality personnel, and administration. If any of the areas for improvement include treatment changes, representatives from the treatment staff and management should also be included.

This chapter (which is based on Renner et al. 1997 and previous versions of this work) outlines a method to identify and prioritize factors limiting the optimized performance of a drinking water distribution system. In addition, tools that can be used to correct the identified limitations are presented. This chapter is divided into discussions of identifying performance-limiting factors, prioritizing those factors, and developing an action plan to improve performance over time.

To work toward optimization of a distribution system, a utility must first be able to identify factors that are contributing to less-than-optimized performance.

Second, the list of identified factors must be prioritized, taking into account what is required to address the factors in terms of impact, urgency, cost, and time to implement a solution. The utility may want to develop two lists: one containing items that can be accomplished in the near term and the other for long-term solutions. In this way, distribution system staff can address these issues initially through solutions that are more easily implemented, while simultaneously working on ones that require a more significant amount of effort, resources, and time. Third, the utility should begin putting together action plans to address the prioritized problems. Fourth, the utility should carry out its action plans, monitoring progress as it goes, and adjusting the plans as needed to ensure successful completion. Throughout this process, the utility should keep detailed records of activities completed and the outcome, modifying future actions, as appropriate, based on evaluation of initial outcomes. Partnership for Safe Water (PSW) subscriber utilities provide regular optimization activity updates as a component of the program's annual reporting process. The steps of the action plan development process are summarized in Table 7-1.

Identification of Performance-Limiting Factors

Throughout this guide, the self-assessment team has been presented with a series of questions that are designed to provide introspective thought on the part of the utility. After carefully considering a question, the utility should decide whether that aspect of the distribution system is Optimized and Documented, Partially Optimized, or Not Optimized. The self-assessment completion report should contain appropriate information and data to adequately support the utility's response to the self-assessment questions. Note that a category is only considered to be optimized if all areas covered in the self-assessment question are optimized—otherwise a status of Partially Optimized or Not Optimized should be selected.

Table 7-9, at the end of the chapter, provides a summary list of the self-assessment questions included in all previous chapters of this guide. These questions are listed in the order in which they appear in this guide. Note that not every self-assessment question will apply to every distribution system (for example, nitrification questions would not apply to a free chlorine system). Distribution systems are only required to address the self-assessment questions that are

Table 7-1.   Steps to action plan development

1. Identify performance-limiting factors
2. Prioritize performance-limiting factors based on impact and urgency
3. Develop an action plan to address performance-limiting factors, with initial consideration of the highest priority items
4. Measure results of actions, make adjustments, reevaluate, and continue the optimization sequence

relevant to their system with regard to their prioritization and potential impact on the performance of the primary optimization parameters. The table in this guide provides space to record the system's optimization status, whereas PSW subscribers may also record information in the Optimization Assessment Tool spreadsheet, provided to program subscribers, which aids utilities in capturing responses and ranking performance-limiting factors. Although the self-assessment questions included in this guide are thorough, they are not comprehensive. The criteria, categories, and factors described in this assessment are among the most common encountered by many systems and have the highest probability of affecting system performance; however, systems are not limited to addressing only the questions in this guide. Additional, system-specific factors may exist that are of great importance to some utilities. These should not be ignored and may be included in the self-assessment process. The optimization status of these factors may be described in the self-assessment completion report, along with any action plans created to address selected parameters, based on their prioritization and potential impact on the performance of the primary optimization parameters.

As part of this process, the assessment team should identify whether or not an item is truly impacting distribution system performance because items can be interrelated, making the root cause of some problems difficult to identify. For example, the root cause of low disinfectant residual values in a chloraminated system may be caused by a variety of factors, including practices at the water treatment plant. Investigation of several factors may be required to identify the root cause of the issue as a change in the ratio of chlorine to ammonia fed at the treatment plant, for example. To discern the root cause of a distribution system issue, at times a special study or series of special studies may need to be conducted, with the objective of proving or disproving the suspected cause of the poor performance.

As special studies are conducted, some items that were previously of concern may be removed from the list, whereas others may be added. An example of the structure of a special study is included in the Appendix of this guide. Establishing buy-in and support from all participants, for the common goal of improving distribution system performance and operations, is critical at this stage. Important, yet challenging to address, factors include administrative issues, such as inadequate funding or lack of communication, which must be addressed with the support of distribution system and overall utility management. Some issues may be resolved quickly, while others may require significant planning and budgetary considerations over a longer period of time.

Areas addressed during the self-assessment process represent areas in which the utility has the opportunity to improve, including: Performance, Design, Operations, and Administration. The detailed questions included in previous chapters of this guide help utilities identify potential areas for improvements. Examples of broad topics addressed in the previous chapters are outlined in the following sections. Utility staff completing the self-assessment completion report should provide detailed narratives that describe the team's rationale and activities as they work to achieve distribution system optimization. Tables 7-4 through 7-8 can help to assist the utility in organizing its responses and formulating a performance improvement action plan.

Prioritization of Performance-Limiting Factors

Once the self-assessment process has been completed, including any special studies and assembly of a final list of performance-limiting factors, the utility then needs to prioritize the list to provide a roadmap to distribution system optimization. Performance-limiting factors should be prioritized in the order of their impact on achieving optimized performance, as well as their urgency. A numerical rating should be assigned to the categories of impact and urgency to quantify the priority of the action items identified during the self-assessment process.

Prioritization of performance-limiting factors is accomplished by a two-step process. First, all factors that have been identified are individually assessed with regard to impact on optimized performance, and they are assigned a 1-5 rating, with 1 equating to minimal impact on performance and 5 equating to the most significant impact on optimized performance. This rating is typically performed for assessment categories receiving a Partially Optimized or Not Optimized rating. Categories that are optimized and documented typically do not result in action items, and are not rated, unless the utility has identified actions that will allow for further improvement in this area.

The second step in prioritizing performance-limiting factors categories that are not fully optimized is to list them in order of urgency, assigning ratings from 1 (least urgent) to 5 (most urgent). List the highest rated Partially Optimized or Not Optimized factor categories in order of assessed urgency. The prioritized summary list of factors provides a valuable reference for the next phase of the assessment, which is developing an action plan and implementing improvements to address the performance-limiting factors and move the system toward optimized performance. An example of the rating scale for prioritizing performance-limiting factors is included in Table 7-2. The Optimization Assessment Tool spreadsheet is a resource provided to Partnership subscribers to assist with the process of ranking performance-limiting factors.

The numerical rating of performance-limiting factors is designed to be a beneficial process for utilities to help guide distribution system optimization activities; however, the resulting ranking may be adjusted based on internal distribution system or utility factors. The key to the assessment is to prioritize the *top* items affecting performance so that clarity is provided to utility personnel implementing optimization actions. In addition, the remaining factors that are not highly rated still represent a significant finding and should be addressed after higher rating items are corrected. These factors may be added to a long-term action implementation plan because of the assessed lower impact of these factors

Table 7-2. Rating scale for prioritizing performance-limiting factors

Rating	Description
5	Major impact on long-term optimization goals, sustained
4	Major impact on short-term optimization goals
3	Important impact on optimization
2	Minor impact but sustained
1	Minor short-term impact

on the performance of the three major optimization criteria. Lower ranking factors that are inexpensive and easy to complete should also be implemented rapidly—do not wait for completion of the self-assessment report to address these items, because ideally they may be addressed and corrected at the time that they are identified. These factors are also a source for providing recognition to utility personnel for adequately addressing those potential sources of problems and making rapid improvements in operations and water quality performance.

As previously mentioned, two lists should be developed: one for short-term actions and the other for longer term, more significant actions. The rationale for this is that if the first item on the list is going to require a great deal of time and/or resources to address, the utility may also simultaneously be making incremental progress on other issues that require less time and resources to correct.

It is also important, when prioritizing the action items, to focus first on the primary performance optimization areas (disinfectant residual, pressure, and main breaks) in need of improvement. All three areas may need improvement to reach optimized performance. In that case, priority may be placed on topics that can influence the performance of multiple criteria (such as flushing). If one of these performance areas is furthest from optimized performance, then topics that can improve this may be selected as the highest priority actions. Design, application of operator concepts, and administration factors often influence all three performance areas. Optimizing these areas can therefore result in an overall performance improvement.

The operational Performance Improvement Variables discussed in Chapter 3 are listed in Table 7-3. This table illustrates the influence of each variable on the performance of each of the three primary optimization areas (disinfectant residual, pressure, and main breaks). When prioritizing optimization efforts pertaining to specific Performance Improvement Variables, it is useful to consider these influences. For example, if performance improvement is desired for disinfectant residual performance, then variables that influence this criterion should be ranked as the highest priority. The table illustrates that some variables have greater influence than others. If these variables are not optimized, addressing these variables may present an opportunity to make a major impact on multiple areas of distribution system performance.

Table 7-3.   Impact of performance improvement variables on primary optimization areas

Performance improvement variable	Optimization areas		
	Disinfectant residual	Pressure	Main breaks
Disinfectant Residual	✓	✓	✓
Cross-Connection Control and Backflow Prevention	✓✓	✓	
Customer Complaints	✓	✓	✓
Disinfection By-Product Control	✓	✓	
Energy Management		✓	
External Corrosion Control			✓✓
Flushing	✓✓	✓	✓
Hydrant, Valve, and Blowoff Maintenance	✓	✓	✓
Internal Corrosion Control	✓		✓✓
Main Breaks		✓	
Nitrification	✓		
Pipeline Installation, Rehabilitation, and Replacement		✓	✓✓
Post-Precipitation, Inorganic Accumulation Control	✓	✓	
Pressure Management	✓	✓	✓✓
Security and Online Monitoring	✓	✓	✓
Storage Tank Operations and Maintenance	✓✓	✓	✓
Water Age Management	✓✓	✓	
Water Loss Control		✓✓	✓
Water Sampling and Response	✓	✓	✓

✓   Influences optimization of performance criteria

✓✓   Major influencing factor on performance criteria

Developing Action Plans

The goal of conducting the self-assessment process and addressing the identified and prioritized performance-limiting factors is to achieve the desired level of performance for an existing water distribution system without making major system modifications. The areas in which performance-limiting factors have been broadly grouped (performance, administration, design, and operation) are all important in that a factor in any one of these areas can individually influence poor performance. However, when a utility is implementing optimization activities, the relationship of these categories to achieving the goal of delivering high-quality water must be understood.

Administration, design, and maintenance-related factors all lead to a system physically capable of achieving optimized performance. It is the operation, or more specifically the system control activities, that enable a capable system to consistently deliver high-quality water. This concept was shown graphically in Figure 1-1. Focusing on system control when implementing optimization activities allows priorities to be developed for making the changes required to improve performance. In this way, the most direct approach to improved performance is implemented.

The prioritized performance-limiting factors should be categorized into the four groups (performance, administration, design, and operations) so that the resources needed to resolve the issue can be assembled. Some issues may not fit neatly into just one group and may cross over into two or more groups. A strong and detailed action plan should contain as many of the following components as possible:

- Definition of the issue
- Identification of the responsible person or department to complete the action
- Required tasks
- Target completion timeline
- Parameters needed to measure success
- Any budget requirements

The Partnership for Safe Water's Phase III Self-Assessment Completion Report Template Document contains blank action plan tables using the format displayed in the following examples that guide utilities to create detailed action plans for distribution system optimization activities. The following examples demonstrate how the resources required to complete an action tend to be group-specific. When assigning tasks to groups, do not overlook the importance of the operations staff in implementing distribution system improvements.

The following are general examples that may not be applicable to all distribution systems. Distribution system staff should develop action plans that are specific to their system and address the performance-limiting factors identified during the self-assessment process.

Example 1 (Performance)

After compiling distribution system disinfectant residual data over the past several years and evaluating the data using a variety of techniques, including mapping, distribution system staff discovered that a specific area of the distribution system was not consistently maintaining adequate free chlorine residual concentrations to meet the PSW's optimization goals. This particular area of the system was constructed in anticipation of a large home development, which was never completed. There are several solutions that the utility could consider to improve disinfectant residual performance in this area. An action plan may contain the short-term action of evaluating these solutions in order to identify the most suitable means to address this issue, as well as perhaps taking steps to implement a temporary solution for maintaining disinfectant residual. The long-term solution may be implementing the selected solution and verifying its success through regular disinfectant residual sampling in this area. This long term-solution likely includes a budgetary component, which should be considered in the action planning process. An example of how this improvement strategy may be expressed in an action plan format is displayed in Table 7-4.

Example 2 (Administrative)

In this example, the utility is unable to complete its scheduled valve exercise program because of inadequate staffing. The utility understands the importance of valve exercising and its impact on distribution system performance; therefore, an action plan was developed that enabled the utility to address both its short-term issue (completing the year's scheduled valve exercising program) and the root cause of the issue (hiring additional staff so that an optimal level of distribution

Table 7-4. Example action plan to address low distribution system disinfectant residuals

Issue	Short-term solution	Person(s) responsible	Target date to be completed	Long-term solution	Person(s) responsible	Target date to be completed
Water quality degradation leads to loss of disinfectant residual in low-use areas of the system.	Send staff to site weekly to test disinfectant residual concentrations and flush as necessary.	Distribution system manager, field staff	Starts immediately, to be conducted until long-term action is completed	Budget for and install chlorine analyzers and auto-flushers at low use/high water-age locations in the system. Continue with regularly scheduled sampling to verify effectiveness.	Distribution system manager, instrumentation technicians, field staff	April 2017

system maintenance could be performed). When developing action plans to address administrative performance-limiting factors, consider that long-term solutions may involve policy or budgetary components. It is important that target completion dates be developed with the consideration of the appropriate timelines for completion of these components. An example of one way this example utility may develop an action plan to address their administrative limitations is displayed in Table 7-5.

Example 3 (Operations)

Although this example utility performs a basic level of pressure monitoring at selected high- and low-pressure sites in the distribution system, a review of existing pressure data indicates that information from additional sites would lead to better understanding of overall system performance, with regard to pressure, the Partnership's indicator of the hydraulic integrity of the distribution system. Both short- and long-term solutions are developed to meet this need. In the short term, portable pressure monitors may be used to obtain pressure data from key distribution system sites, whereas the system's hydraulic model may be applied to assist in the identification of additional pressure monitoring sites throughout the system, taking into account potential power and communications needs. Longer term solutions may include budgeting for pressure monitors, installing the monitors, setting up any required communication mechanisms, and developing a plan to regularly evaluate and respond to information generated by the pressure monitors. An example of how this improvement strategy may be expressed in an action plan format is displayed in Table 7-6.

Table 7-5.   Example action plan to address staffing issues

Issue	Short-term solution	Person(s) responsible	Target date to be completed	Long-term solution	Person(s) responsible	Target date to be completed
Inadequate staffing results in the inability to complete valve exercise according to the utility's preferred 3-year schedule.	Hire outside support to assist in performing valve exercises for critical valves for this year.	Operations manager	December 2015	Budget for additional maintenance staff. Hire and train additional maintenance staff that can support a sustainable valve exercise program.	Operations manager, utility manager	December 2016

Example 4 (Design)

Asset management plays an important role in distribution system optimization and can have a significant impact on the system's water quality integrity as well the status of the system's physical components. In this example, a utility regularly performs valve exercises, but the information obtained from the process is not integrated into the utility's asset management system. Because of this, the information generated during valve exercise is not regularly evaluated or considered fully in utility planning for asset rehabilitation and replacement. Both short- and long-term actions were developed to address this performance-limiting factor. In the short term, because valve exercise information was being recorded, the utility planned to continue recording the relevant information in a separate file, whereas long-term plans were developed to integrate valve exercise information into the utility's asset management system as well as regularly evaluate the information to help support decisions related to the physical status of the system's valves. An example of how this improvement strategy may be expressed in an action plan format is displayed in Table 7-7.

Example 5 (All of the Above)

As described in chapter 6, changes in a utility's raw water source have the potential to significantly impact water quality throughout the water treatment plant

Table 7-6. Example action plan to address pressure monitoring needs

Issue	Short-term solution	Person(s) responsible	Target date to be completed	Long-term solution	Person(s) responsible	Target date to be completed
Additional pressure monitoring is required to obtain accurate pressure information from the distribution system.	Use portable/temporary pressure monitors to obtain data for critical sites. Use hydraulic model information to determine the optimal sites for future pressure monitoring.	Distribution system manager, field staff	Starts immediately, to be conducted until long-term action is completed.	Budget for and install pressure monitors for critical sites in the system. Develop a plan to evaluate data from the pressure sensors.	Distribution system manager, instrumentation technicians, field staff	April 2017

Table 7-7.   Example action plan to address pressure monitoring needs

Issue	Short-term solution	Person(s) responsible	Target date to be completed	Long-term solution	Person(s) responsible	Target date to be completed
Information from valve exercise is not currently not recorded in the utility's asset management database, so it is not accessed regularly or used for decision-making.	Continue to record valve exercise in separate file.	Field operations and maintenance staff, information technology staff	Ongoing until long-term solution is implemented.	Develop a module for the asset management program so that valve exercise information can be incorporated. Train staff on how to record information. Develop a schedule to periodically review the information.	Distribution operations and maintenance manager, field staff, information technology staff	December 2016

and distribution system. Because of this, it is important that utilities fully evaluate and carefully consider the potential impact of such a change before its implementation. In this example, a utility considers bringing a new raw water source online to address water rights issues. Table 7-8 provides an example of an action plan designed to address considerations related to this scenario. It includes short-term actions related to policy as well as more technical actions related to characterization of the potential new water source and bench/pilot-scale testing of treatment and disinfection processes using the new water source. Longer term actions pertain to evaluation of pilot results to determine the feasibility of implementing the new water source, seeking appropriate regulatory approvals, developing a plan to bring the new water source into service (if appropriate), and developing water quality targets and protocols for testing. An example of how this may be expressed in an action plan format is provided in Table 7-8. Although this table addresses many concerns related to source water changes, it is not intended to be comprehensive. It is suggested that utilities considering a source water change work with their local regulatory agency to help ensure that all system-specific water quality concerns and regulatory requirements are addressed.

Table 7-8.   Example action plan to address potential water source changes

Issue	Short-term solution	Person(s) responsible	Target date to be completed	Long-term solution	Person(s) responsible	Target date to be completed
Utility is considering bringing a new water source online because of water rights issues and needs to consider the impact of the change on distribution system water quality.	Review and update utility policy regarding modifications in raw water source. Fully characterize raw water quality, considering seasonal variations. Perform bench- and pilot-scale testing of treatment processes, including corrosion impacts.	Utilities manager, Engineering manager, Water treatment staff, Water quality laboratory staff	December 2016	Determine whether pilot results support use of the additional raw water source. Develop a plan to bring the source online if appropriate. Seek appropriate regulatory approvals. Set performance goals and testing schedule for use of new water source.	Utilities manager, engineering manager, water treatment staff, distribution system staff, water quality laboratory staff	December 2017

Implementing Action Plans

At this point in the self-assessment process, the utility should have a prioritized list of actions to complete that includes:

- Identification of the issues being addressed
- Specific actions and tasks
- Required resources
- Target completion date and/or timelines
- Factors that determine success
- Responsible individual(s) or groups

As action plans are implemented, the utility needs to periodically ask the questions "Are we headed in the right direction and is progress occurring at the expected pace?" If the answer to either of these questions is no, the utility needs to reassess the approach it is taking to solve the problem and, potentially, modify the action in order to achieve the desired outcome. The change could be relatively minor (such as a modification to the target completion date) or very significant, based on evaluation of the goals, progress, and timeline. Utilities and distribution systems can and do change, and the action plans developed during the self-assessment process should change with them, as appropriate. Establishment of periodic check-in points for action plans is critical because they provide opportunities for critical evaluation of the progress being made. At these points, individuals should be involved in the progress assessment who are objective and can provide constructive criticism when appropriate.

When implementing action plans, some action items can be a long-term process, requiring a long-term commitment for the utility. The rationale is described here:

- *Greater Effectiveness of Training to Accomplish Skills Transfer.* Operator and administrator training can be conducted under a variety of actual operating conditions (for example, seasonal water quality and demand changes). This approach allows transfer of the skills necessary to maintain high quality water even during periods when the system is under stress. Training may need to be conducted over an extended period to encounter many operating conditions.

- *Time Required to Make Administrative Changes.* Administrative changes often require an extended period to address. For example, if the utility rate structure is inadequate to support optimized system performance, extensive time can be spent developing required changes in the rate structure and gaining political support to implement them. Communication barriers between labor and management may have to be addressed for improved performance. Development of new utility policy, or the modification of existing policy, can require time to secure the necessary approval from all stakeholder groups.

- *Time Required to Make System Modifications or to Implement Deferred Maintenance Activities.* For changes requiring financial expenditures, the necessary time and a multiple-step approache are typically required to gain administrative approval. First, the need for system modifications or deferred maintenance improvements must be demonstrated through system control efforts. Then, administrators must be shown the need and ultimately convinced to approve funds necessary for the improvements. These activities can take considerable time before the identified modification or deferred maintenance item can be approved and corrected.

Experience has shown that no single approach can address the unique combination of factors in every water system; therefore, the actual details of implementation are system-specific and should be completed by utility personnel. Selected tools to aid in action plan development and implementation, such as a special study, standard operating procedures (SOPs), system control data sheets, and data collection and trending software, are described in greater detail in the Appendix.

The optimization process is a continuous process because utilities are encouraged to measure progress, reassess performance, and develop new and relevant optimization actions on an ongoing basis. This process is reinforced through the PSW's annual reporting process.

Conclusion—Optimization and Continuous Improvement

Congratulations on taking the first important step on the journey to distribution system optimization—performing a self-assessment! Utility personnel have undertaken a critical evaluation of the distribution system and how it operates. Challenges have been identified that are limiting performance and the ability to achieve optimization. Now the work begins.

Refer to the action plan tables developed from the self-assessment findings from the distribution system and begin the journey. The value of the self-assessment will continue to reveal itself as operations team members continue to pursue excellence in distribution system operations in the days, weeks, and years to come. The strength of the self-assessment lies in understanding the details that enable the system to deliver water of the highest quality to its customers, while meeting its additional objectives. As one of the early contributors to the Partnership, Robert Renner, shared:

"Operational excellence is doing simple things exceedingly well."

The Partnership challenges operators who have participated in a self-assessment to be deliberate at moving to the next step of optimization—addressing the performance-limiting factors identified in the self-assessment and being tenacious at completing the action plans developed. As operators and operations team members complete action plans and performance improvements are realized, momentum will build and a culture of excellence will begin to permeate the utility. At this stage, it is critical to "fan the flame" and reinforce the positive steps completed while continuing to encourage and support future plans for improvement. The journey to operational excellence will require tenacity, patience, perseverance, communication, and the commitment of all operations team members. The journey is long and the end goal of excellence in optimization is truly never reached—there will always be additional improvements that can be implemented and additional performance gains that can be achieved. In the end, the lessons learned during the journey of striving for excellence are the reason for taking the first step.

Table 7-9. Comprehensive summary of self-assessment questions

| Self-assessment category | Questions for gauging optimization status | Optimization status | | | Comments |
		Optimized and documented	Partially optimized	Not optimized	
Chapter 2—Performance Assessment					
Disinfectant Residual	Do the disinfectant residual data meet the optimization performance goals? The goals are listed here and apply to both the entire distribution system and to each sampling location. Disinfectant residual in 95% of the monthly and yearly routine measurements: • Free chlorine - ≥ 0.20 mg/L and ≤ 4.0 mg/L • Total chlorine - ≥ 0.50 mg/L and ≤ 4.0 mg/L • Chlorine dioxide - ≥ 0.20 mg/L and ≤ 0.80 mg/L				Submit data collection spreadsheet.
	Are there any consecutive disinfectant residual measurements from optimized sample sites below the disinfectant residual goals?				
	Does the utility have a system sampling map and schedule? If so: • Is utility staff sampling at the (previously described optimized sampling locations)? • Are sampling locations representative of distribution system water quality? • Are sampling locations tracked that repeatedly exhibit low disinfectant residuals? • Are nonroutine low residual sample sites added to the future routine sampling schedule? • Are performance improvement variables (chapter 3) used to reduce low residual recurrence?				

Self-assessment category	Questions for gauging optimization status	Optimization status			Comments
		Optimized and documented	Partially optimized	Not optimized	
Disinfectant Residual (*continued*)	Is disinfectant residual testing performed using approved methods and digital testing equipment? Are values recorded to two decimal places, where possible?				
	Are there online continuous chlorine analyzers in use throughout the distribution system? Are data from these analyzers collected and continuously displayed for operators by the supervisory control and data acquisition (SCADA) system?				
Pressure Management	Does the system meet all of the pressure management optimization goals? • 20 psi minimum (under normal operating conditions including maximum hourly demand and fire flow) in 99.5% of the daily minimum measurements • Maximum pressure does not exceed utility specified maximum in 95% of measurements • Pressure fluctuations do not exceed utility specified maximum pressure fluctuation range in 95% of measurements				Submit pressure data collection spreadsheet. Report utility specified goals for pressure maximum and range.
	Is pressure monitored at a minimum of two critical sites in each pressure zone (areas of high and low pressure)? Is the pressure at these sites monitored continuously? Are the instruments properly installed and routinely calibrated?				
	Does pressure monitoring include maximum day demand flow, fire flow events, and emergency situations (such as a main break or power outage)?				
	If online pressure monitoring is available, is the data analysis configured to alarm the operator when low pressure (< 35 and/or 20 psi) or high pressure spikes occur?				

Self-assessment category	Questions for gauging optimization status	Optimization status			Comments
		Optimized and documented	Partially optimized	Not optimized	
Pressure Management (*continued*)	Are pressure fluctuations monitored and investigated, and are procedures used to reduce the range and duration of pressure variations?				
	Are there preapproved procedures (written SOPs) that help prevent low-pressure events and to protect public health in case of pipeline depressurization (from main breaks or due to power outages)?				
	Is the operator aware of conditions that can cause low or high system pressures? Are SOPs available that account for routine and nonroutine operations that might affect pressure? Are maximum system pressures and pressure fluctuations documented?				
Main Breaks	Does the system meet the main break goals?				Submit main break frequency data collection spreadsheet.
	Are main breaks correlated to variations in pressure? Is there an opportunity to reduce system pressures while still meeting all pressure management goals to reduce the frequency of main breaks?				
	Has the utility established SOPs to ensure appropriate and timely response to main break/depressurization events? Is all appropriate staff trained in these procedures? Are SOPs regularly reviewed and updated?				

Self-assessment category	Questions for gauging optimization status	Optimization status			Comments
		Optimized and documented	Partially optimized	Not optimized	
Chapter 3—Performance Improvement Variables					
Cross-Connection Control and Backflow Prevention	Does the utility have authority to enforce backflow regulations (local, state/provincial, federal, or other)?				
	Does the utility have a comprehensive list of locations where backflow prevention devices are required and testing and verification schedules and results?				
	Does the utility provide training and verify certification for cross-connection installers and testers?				
	Does the utility have a comprehensive backflow prevention program that includes all the elements in AWWA Manual M14 and meets all local, state/provincial, or other applicable requirements?				
Customer Complaints	Are technical customer complaints recorded and tracked separately from billing and general information inquiries?				
	Is customer water quality complaint response time tracked?				
	Does the number of annual technical water quality complaints compare favorably with the AWWA benchmarking survey results (2.5 per 1000 customer accounts per year)? If not, is there a plan to reduce this number?				Calculate and report the number of technical complaints/1000 customer accounts.
Disinfection By-Products	Do TTHM and HAA5 distribution system test results satisfy the regulatory requirements? (USEPA maximum contaminant levels are 80 µg/L for TTHM and 60 µg/L for HAA5, calculated as a locational running annual average—other countries may vary)				
	Are system DBP concentration trends monitored, and is there a plan to maintain or reduce the current levels while maintaining adequate disinfectant residual?				Assess the results shown on the disinfection performance spreadsheet.

Self-assessment category	Questions for gauging optimization status	Optimization status			Comments
		Optimized and documented	Partially optimized	Not optimized	
Energy Management	If pressure is managed to reduce energy usage does the utility monitor the effect on stability insuring that appropriate pressure ranges are maintained?				
	Are all major distribution system pumps routinely tested for efficiency? Are there targets for maintenance or replacement based on efficiency?				
	Has a hydraulic surge analysis been performed and addressed where required?				
External Corrosion Control	Are pipes inspected and sampled whenever they are exposed?				
	Are corrosion observations recorded during all main break and leak repairs?				
	Are soils always tested before installing metal pipes or materials?				
	Is corrosion protection installed when materials and environmental conditions warrant such measures?				
Flushing	Has the utility instituted a routine flushing program and documented rationale for flushing practices (or lack of) currently in use?				
	Are the following methods incorporated into the flushing program? • Is chlorine residual monitored as part of the flushing program? • Is chlorine residual used as an indicator of when it is acceptable to terminate a flush or used to adjust automatic flushing settings as appropriate? • Are routine flows or flush velocities at least 2.5 ft/sec when removing loose particles? • Are routine flushing velocities at least 5 ft/sec for removal of cohesive and loosely adhered particles and biofilm? • Is dechlorination of flushing water provided where appropriate?				

Self-assessment category	Questions for gauging optimization status	Optimization status			Comments
		Optimized and documented	Partially optimized	Not optimized	
Flushing (*continued*)	Is pressure monitored during flushing (manual and automatic) to verify that pressure goals are continuously achieved?				
Maintaining Hydrants, Valves, and Blowoffs	Does the system have accurate and current records that document the location and attributes for all valves, hydrants, and blowoffs?				
	Are all valves, hydrants, and blowoffs inspected and evaluated on a schedule?				
	Are all distribution system main valves and hydrants exercised and tested at least every three years (or more frequently if required by regulation)?				Assess the valve exercising program to ensure that the testing frequency goal is met.
	Are all hydrant repairs scheduled within 24 hours of discovery? Are inoperable hydrants identified immediately and is this communicated to the fire protection authority?				
	Does the system control access to hydrants and provide training for proper third-party use?				
Internal Corrosion	Does the system meet regulatory lead and copper action levels?				
	Does the utility have a corrosion testing strategy (such as coupons, electronic detection, water quality monitoring) conducted routinely at locations throughout the system?				
	Is the break site examined for tuberculation, pitting, holes, or scaling?				
	Does the utility have a program established to address potential impacts on corrosion when changing source water and/or treatment conditions?				
Nitrification	Are free ammonia, nitrite, and heterotrophic plate count (HPC) tested routinely? Are action levels established that are based on test results?				Assess utility action level goals for these parameters.

Self-assessment category	Questions for gauging optimization status	Optimization status			Comments
		Optimized and documented	Partially optimized	Not optimized	
Nitrification (*continued*)	Is the total chlorine residual maintained at a concentration greater than 0.50 mg/L? Are storage tanks monitored, particularly in areas lacking circulation? Are zone boundaries and dead ends monitored for nitrification?				Use disinfectant residual performance spreadsheet to assist with this evaluation.
Pipeline Installation, Rehabilitation, and Replacement	Are all pipelines installed and disinfected as required by applicable regulations? Do procedures follow AWWA/ANSI Standards C600-620 and C651 or others, as appropriate?				
	Is there a formal process to prioritize main replacements? Is the rate of replacement and rehabilitation adequate to reduce the amount of unlined metal pipe in the system?				
	Is the renewal rate adequate to reduce the pipe mileage that has the highest risk of failure?				
Post-precipitation, Inorganic Accumulation	Are areas of low disinfectant residual investigated for precipitation?				
	Are treatment plant practices optimized to reduce precipitation potential?				
	Are areas of precipitation accumulation cleaned or flushed at least annually (storage tanks, low water use areas, dead-end mains)?				Evaluate the plan to reduce or eliminate continued precipitation.
Security and Online Monitoring	Have the items identified in the distribution system vulnerability assessment/analysis been implemented?				
	Is there an emergency response plan? Are plan exercises conducted regularly? Are items identified during an exercise followed up?				Evaluate the frequency and effectiveness of exercises.
	Are disinfectant residual records readily available for reference during an emergency?				

Self-assessment category	Questions for gauging optimization status	Optimization status			
		Optimized and documented	Partially optimized	Not optimized	Comments
Storage Facility Operation and Maintenance	Are storage facilities routinely (at least weekly) sampled for disinfectant residual testing? Are special sampling surveys conducted periodically to assess residual uniformity?				
	Is water use from the storage facility monitored? Is water age in the storage facility tracked and controlled? Is the tank turnover rate optimized for water age consideration?				
	Is a routine inspection of all storage facilities conducted annually or according to regulatory requirements (if more frequent)? Are formal procedures for routine inspections in place and required maintenance identified and addressed?				Routine inspections are described previously and are mainly external.
	Are all storage facilities cleaned at least every 5 years? Is service interrupted when facilities are isolated for cleaning?				Evaluate cleaning frequency depending on inspection results.
Water Age, Modeling	Does the system have known areas where water age is high? Are these areas closely monitored for disinfectant residual, microbial parameters, and DBPs?				
	Is water age monitored using a calibrated hydraulic model and current operating conditions? Is the system operated to minimize water age?				Assess the use of the hydraulic model to identify excessive water age areas and situation.
	Does the system have a calibrated hydraulic model that includes water quality parameters? Is the model used regularly and recalibrated as changes are made?				

Self-assessment category	Questions for gauging optimization status	Optimization status			Comments
		Optimized and documented	Partially optimized	Not optimized	
Water Loss Control	Does the system use the AWWA/International Water Association (IWA) water audit method described in AWWA Manual M36? Does the system calculate and track the Infrastructure Condition Factor, the Infrastructure Leakage Index (ILI), and the annual real losses? Is a system water audit performed annually? If so, is the water audit third-party validated?				Calculate the ICF, ILI, and real loss volume and compare to past results. Review water loss procedures to ensure they conform to M36.
	Is the system divided into DMAs or pressure zones to optimize leak detection? Does the system have an active leak detection program (acoustic or other)?				
	Does the utility have a defined meter replacement schedule? Is meter accuracy verified before new meters are installed? Are meters removed from use tested to verify that the meter replacement schedule is effective? If advanced metering infrastructure (AMI) is in place, is it used to notify consumers of possible private leaks?				
Water Quality Sampling and Response	Is there a routine sampling and testing plan that includes monitoring beyond regulatory requirements? Does the plan address system-specific issues?				
	Are routine samples (at least weekly) taken from all storage tanks (and facilities) and tested for microbial parameters, disinfectant residual, and other appropriate parameters?				
	Is an SOP available and followed for main break or leak repair, including any required sampling and testing?				Review the main break testing procedures to ensure they are followed.

Self-assessment category	Questions for gauging optimization status	Optimization status			Comments
		Optimized and documented	Partially optimized	Not optimized	
Chapter 4—Design Evaluation					
Distribution System Design Tools	Does the utility have a distribution system hydraulic model?				
	Does the hydraulic model incorporate the following characteristics: • Water age information • Disinfectant residual (chlorine or chloramine) decay information				
	Is the output from the model compared with real system data for validation purposes?				
	Does a process exist to periodically update the hydraulic model to reflect the dynamic nature of the distribution system?				
Asset Management	Does the distribution system have an up-to-date schematic map that indicates all major physical assets including distribution mains, storage facilities, pumps, hydrants, valves, and meters?				
	Does a complete inventory of all distribution system assets exist?				
	Are maintenance records kept that include maintenance procedures performed, maintenance dates, and asset condition observations made at the time maintenance is performed?				
	Does an asset management program exist that incorporates the following elements? • Condition assessment • Residual life determination • Replacement cost estimate • Established level of service targets • Criticality determination • Development of a capital improvement plan?				

Self-assessment category	Questions for gauging optimization status	Optimization status			Comments
		Optimized and documented	Partially optimized	Not optimized	
Pipeline Materials	Is there a complete distribution system pipeline inventory that includes information about pipeline size, length, age, location, and materials of construction?				Describe criteria used to prioritize replacement and repair.
	Does the utility's asset management plan include specific criteria that are used to determine pipeline replacement?				
	Has the utility determined the total length and materials of construction for distribution system pipeline that is more than 75 years old?				
	Has the utility identified all areas in which unlined metal pipe is present in the system?				
Storage Facilities—Materials and Construction	Has the size, age, and materials of construction been identified for all distribution system storage tanks?				Report tanks that are currently scheduled for replacement or repair.
	Has a condition assessment been performed for all storage facilities?				
	If the system contains tanks that are greater than 30 years old, has a rehabilitation or replacement plan been developed to address these tanks?				
Pumping Facilities	Does the utility have an inventory of distribution system pumping facilities, including information related to capacity, number and size of pumps, age of pumps/installation date, inspection results, and strategy for redundancy?				Report pumps that are currently scheduled for replacement or repair.
	Has the utility developed standard operating procedures (SOPs) for pumping facility operation and maintenance that include predictive and preventive maintenance elements?				

Self-assessment category	Questions for gauging optimization status	Optimization status			Comments
		Optimized and documented	Partially optimized	Not optimized	
Pumping Facilities (*continued*)	Is the Planned Maintenance Ratio calculated and tracked for pumping facilities?				Report the Planned Maintenance Ratio for the 12-month reporting period. Describe the plan to improve this performance measure.
	Does the asset management plan include specific criteria for replacement and rehabilitation of pumps in its pump replacement/rehabilitation strategy?				
	Does the system have adequate back-up pumping capacity to have the firm capacity to meet the maximum date demand and a continuous minimum pressure > 20 psi?				
Valves and Hydrants	Does the utility have a current (updated) valve and hydrant inventory? Do valve and hydrant records include installation date and inspection results?				Report the date of the last inventory update.
	Does the utility's asset management plan include specific criteria for replacement and rehabilitation of valves and hydrants?				
Chapter 5—Application of Operational Concepts					
Water Quality Control Testing	Does the utility have a routine distribution system water quality testing program, including established water quality goals and sampling schedules?				
	Are test results communicated to distribution system operators?				
	Do operators make adjustments to the appropriate system controls based on the water quality test results?				

Self-assessment category	Questions for gauging optimization status	Optimization status			Comments
		Optimized and documented	Partially optimized	Not optimized	
Water Quality Control Testing (*continued*)	Has all of the operational staff been involved in the development of the system control program and SOPs), including emergency response guidelines?				
Recognizing Performance Deviations	Does operations staff review data frequently enough to maintain water quality within optimization ranges established and recognize performance deviations?				
	Are spreadsheets, tables, or other tools used by operators to turn data into information that can be used to make process control decisions? Are data summarized with charts and tables, available to all appropriate staff for review? Is a distribution system team assembled to review data periodically?				
	Do operators receive training on data review and analysis that includes the following topics?— • Access to appropriate data outputs • Use real data to recognize changes that may require investigation • Highlight any situations in which water quality deterioration was averted • Data invalidation protocol				
System Evaluation Response	Do all operating staff have the following characteristics? • The tenacity to achieve the performance goals • A willingness to take responsibility for system performance. • A willingness to learn. • Confidence to make changes. • Empowerment to make changes. • Understanding of when to call for help and whom to call.				
	Is the operating staff able to communicate their concern regarding water quality deterioration, in particular the disinfectant residual level, to other departments?				

Self-assessment category	Questions for gauging optimization status	Optimization status			Comments
		Optimized and documented	Partially optimized	Not optimized	
System Evaluation Response (*continued*)	Does utility staff, working as a team, practice "what-if" thinking to develop contingency plans for responding to abnormal or atypical distribution system situations?				
	Can the appropriate operating staff conduct the necessary hydraulic model simulations to determine the most appropriate response to a given situation—or does the staff have the ability to contact a party that does?				
	Are water quality data shared with system operators so that they can make operational adjustments that result in improvement?				
	Does appropriate staff have the knowledge to maintain process components (pumps, valves, etc.) and understand the tasks of a preventive maintenance program?				
	Is the distribution system staff evaluating appropriate parameters, including daily demand, to help determine and control water age?				
	Do operators review their procedures to ensure that long-term performance objectives are included?				
Communication	Has a formal communication protocol for all critical distribution system staff been developed?				
	Do system operators from various shifts and disciplines effectively communicate with each other?				
	Do system operators and maintenance workers effectively communicate to ensure distribution system maintenance status is provided to all system operators frequently and accurately?				
	Is information communication encouraged between treatment plant staff and distribution system staff?				
	Does the utility ensure that employees understand the information presented?				

Self-assessment category	Questions for gauging optimization status	Optimization status			Comments
		Optimized and documented	Partially optimized	Not optimized	
Communication (*continued*)	Does utility leadership ensure that subordinate employees receive and understand the information in a timely manner, especially distribution system operators who spend the majority of their time in the field?				
Online Instrumentation and SCADA	Has a maintenance and calibration schedule been implemented for all instrumentation?				
	Has the utility implemented a training program to train system operators or maintenance technicians in the care and maintenance of on-line water quality instrumentation? Has the utility implemented a training program to train system operators in the use and operation of the SCADA system and field SCADA interfaces?				
	Does the utility have clearly defined roles on who is permitted to modify and maintain the SCADA system and are those individuals properly trained?				
	Application and placement—Are the instruments selected suitable for the application and physical placement for which they are installed? Does the placement of the instruments allow for ease of access for maintenance? Are environmental conditions suitable for analyzer placement? Is the placement of the sample collection and delivery lines suitable to obtain a representative sample?				
	Calibration/verification—are the instruments calibrated and verified regularly? Do system operators follow the manufacturer's instructions for calibration and verification? Are calibration and verification performed at a frequency that, at minimum, meets regulatory requirements? Are calibration records maintained and accessible when needed? Has an SOP been developed for the calibration and maintenance of all critical instrumentation?				

Self-assessment category	Questions for gauging optimization status	Optimization status			
		Optimized and documented	Partially optimized	Not optimized	Comments
Online Instrumentation and SCADA (*continued*)	Are control systems tested and checked regularly to ensure proper operation and functionality when needed? This may include testing functionality of alarms and control capabilities.				
	Redundancy—are redundant instruments and equipment installed where appropriate? Does the system have a plan established for the procurement of all critical parts to move toward making the distribution system "bulletproof"?				
	Does the utility have provision for a backup server or SCADA system in the event of communication loss or failure? Does the distribution system operator have adequate quantity of equipment to monitor SCADA from the field? Is there an SOP to manually operate the distribution system in the absence of SCADA or a prolonged power outage?				
	Is all pertinent information documented? Is record-keeping adequate (i.e., is the necessary information maintained for the amount of time required by regulatory parties or internal system processes)?				
	Do distribution system operators have adequate capability to view critical system data while working in the field?				
	Does the control system provide operational "safety nets," consistent with system practices and process performance goals, to protect the operator and water quality?				

Self-assessment category	Questions for gauging optimization status	Optimization status			Comments
		Optimized and documented	Partially optimized	Not optimized	
Chapter 6—Administration					
Administrative Policies	Do the utility's overall goals and commitment, at the highest level of management indicate a focus on supplying the highest quality of water possible to customers, and include the role of distribution system operators in achieving these goals?				
	Does the utility have a strategic planning process?				
	Does operating staff have authority to make required operation and maintenance decisions?				
	Do management styles, organizational capabilities, or communication practices at any management level adversely affect performance?				
	Do administrators have a firsthand knowledge of distribution system needs, through site visits or discussions with utility staff?				
	Does the utility have ongoing public information activities, including communications with customers on water quality issues and improvements?				
	Has the utility maintained awareness of existing and impending water quality regulations, and is this information used to develop long-term plans to ensure compliance?				
	Does management support utility staff involvement in professional organizations?				
	Does management prioritize water quality over water quantity?				

Self-assessment category	Questions for gauging optimization status	Optimization status			Comments
		Optimized and documented	Partially optimized	Not optimized	
Acceptance of Optimization Goals	Does operational guidance reflect the optimization goals adopted by the utility?				
	Does operations staff maintain a focus on goals even during challenging water quality and/or operational events that require changes to specific operating parameters?				
	Does administrative staff embrace and pursue goals "beyond the regulations"?				
	Does the organization's mission statement embody the Partnership philosophies?				
	Is distribution staff empowered to make operating changes to address potential water quality issues, or are proposed changes required to be approved by management?				
	Can appropriate changes be made during the evenings, weekends, and holidays?				
Involvement of All Parties in the Partnership Process	Has the organization developed a vision and/or mission statement that incorporates the PSW process?				
	Is there strong leadership throughout the organization that supports the PSW culture?				
	Have all personnel been briefed on the purpose of the Partnership for Safe Water Program and do they have a good understanding of the overall intent to optimize distribution system operation, as well as the benefits of being a member of the Partnership for Safe Water?				
	Does each employee feel that his or her role is well defined and do employees understand their roles in the optimization process?				
	Do administrators set goals and provide an environment that expects and encourages involvement in optimization at all levels?				

Self-assessment category	Questions for gauging optimization status	Optimization status			Comments
		Optimized and documented	Partially optimized	Not optimized	
Documentation/Demonstration of Addressing Complacency	Does the utility employ the best practices listed in Table 6-1 to combat complacency?				
	During the past 10 years, how has staff responded to a major episode of diminished water quality in the distribution system?				Provide description.
	If complacency was determined to be the cause of a water quality episode, what actions have been taken to prevent it from occurring in the future? What process control procedures and special studies were completed related to the identified cause of complacency?				Provide description.
	How would staff respond to deteriorating water quality during unusual changes in water quality in the distribution system?				Provide description.
	How would staff address a situation in which all of the usual operational control procedures do not maintain optimized performance?				Provide description.
	Does the utility have a policy/procedure to implement when considering significant changes to the type of source water or water quality entering the distribution system, thatconsiders the impact of water quality changes on overall distribution system water quality?				
Training	Does training/documentation exist for new employees?				
	Does training/documentation exist for all employees?				
	Does the utility ensure that employees receive appropriate safety training?				
	Is the distribution system staff knowledgeable about drinking water regulations and emerging technologies?				
	Does a culture of learning and continuous improvement exist throughout all levels of staffing?				

Self-assessment category	Questions for gauging optimization status	Optimization status			
		Optimized and documented	Partially optimized	Not optimized	Comments
Staffing	Are crews of sufficient size to accomplish necessary maintenance and repair activities in a safe and timely manner?				
	Are response time goals (state goals) consistently met?				
	Is the number of distribution system staff adequate to maintain optimized performance?				
	Do all distribution system operators possess the appropriate level of certification/licenses to operate the distribution system?				
	Does the utility have the instrumentation to continuously monitor all critical parameters with alarms?				
	Does the current staffing level have a detrimental effect on distribution system operation, maintenance, or sampling procedures?				
	Does a high staff turnover rate, high degree of absenteeism, or large number of grievances indicate other underlying staffing problems?				
	Does the improper distribution of adequate staffing result in poor distribution system performance or response to performance deviations?				
	Does a low pay scale or benefit package discourage more highly qualified personnel from applying for utility positions or cause personnel to leave once they are trained?				

Self-assessment category	Questions for gauging optimization status	Optimization status			Comments
		Optimized and documented	Partially optimized	Not optimized	
Funding	Does the utility have the financial health to be self-sufficient?				
	Do revenues cover expenses, and is there sufficient funding to cover unexpected expenditures?				
	Review all major expenditures over the past 5 years and planned future expenditures. Do past and planned expenditures support the utility's water quality goals?				
	Does the utility's bond indebtedness limit funds available for other needed items?				
	Is a declining population or water use trends expected to reduce water usage and revenues?				
	Does the utility calculate values for performance indicators? Does the utility perform comparisons to external financial benchmarks?				
	Are utility rates sufficient to recover operations and maintenance expenses (minus depreciation expense) as well as provide for a healthy capital investment/reinvestment including debt service?				
	Does the system have reserves to cover 6 months of operations and maintenance expenses?				

Definitions—DBP, disinfection by-product; HAA5, sum of five haloacetic acids; SCADA, supervisory control and data acquisition; SOP, standard operating procedure; TTHM, total trihalomethane.

REFERENCES

Antoun, E.N.; Dyksen, J.E.; & Hiltebrand, D.J., 1999. Unidirectional Flushing: A Powerful Tool. *Journal AWWA,* 91:7:62.

ASTM, 2015. D2688—15 Standard Test Methods for Corrosivity of Water in the Absence of Heat Transfer (Weight Loss Methods). West Conshohocken, Paenn.: ASTM.

ASTM, 2013. G96—90 Standard Guide for Online Monitoring of Corrosion in Plant Equipment (Electrical and Electrochemical Methods). West Conshohocken, Paenn.: ASTM.

AWWA/EES (Economic and Engineering Services), 2002a. *Effects of Water Age on Distribution System Water Quality.*

AWWA/EES, 2002b. *Nitrification.*

AWWA/EES, 2002c. *Finished Water Storage Facilities.*

AWWA, 2007. ANSI/AWWA C510-07, Double-Check Valve Backflow-Prevention Assembly. Denver, Colo.: AWWA, Denver.

AWWA, 2015. ANSI/AWWA C512-15, Air-Release, Air/Vacuum, and Combination Air Valves for Water and Wastewater Service. Denver, Colo.: AWWA, Denver.

AWWA, 2010. ANSI/AWWA C600-10, Installation of Ductile Iron Water Mains and Their Appurtenances. Denver, Colo.: AWWA, Denver.

AWWA, 2011. ANSI/AWWA C602-11, Cement-Mortar Lining of Water Pipelines in Place—4 in. (100 mm) and Larger. Denver, Colo.: AWWA, Denver.

AWWA, 2013. ANSI/AWWA C605-13, Underground Installation of Polyvinyl Chloride (PVC) and Molecularly Oriented Polyvinyl Chloride (PVCO) Pressure Pipe and Fittings. Denver, Colo.: AWWA, Denver.

AWWA, 2014. ANSI/AWWA C651-14, Disinfecting Water Mains. Denver, Colo.: AWWA, Denver.

AWWA, 2011. ANSI/AWWA C652-11, Disinfection of Water Storage Facilities. Denver, Colo.: AWWA, Denver.

AWWA, 2015. ANSI/AWWA C670-15, Online Chlorine Analyzer Operation and Maintenance. Denver, Colo.: AWWA, Denver.

AWWA, 2014. ANSI/AWWA D102-14, Coating Steel Water Storage Tanks. Denver, Colo.: AWWA, Denver.

AWWA, 2009. ANSI/AWWA D103-09, Factory-Coated Bolted Carbon Steel Tanks for Water Storage. Denver, Colo.: AWWA, Denver.

AWWA, 2011. ANSI/AWWA D104-11, Automatically Controlled, Impressed-Current Cathodic Protection for the Interior Submerged Surfaces of Steel Water Storage Tanks. Denver, Colo.: AWWA, Denver.

AWWA, 2010. ANSI/AWWA J100-10 (R13), Risk and Resilience Management of Water and Wastewater Systems (RAMCAP). Denver, Colo.: AWWA, Denver.

AWWA Disinfection Systems Committee, 2008a. Committee Report: Disinfection Survey, Part 1—Recent changes, current practices, and water quality. *Journal AWWA*, 100:10:76.

AWWA Disinfection Systems Committee, 2008b. Committee Report: Disinfection Survey, Part 2—Alternatives, experiences and rationales. *Journal AWWA*, 100:11:110.

AWWA, 2015. G200-15, Distribution System Operation and Management. Denver, Colo.: AWWA, Denver.

AWWA, 2016a. M17—*Fire Hydrants: Installation, Field Testing, and Maintenance*, 5th ed. Denver, Colo.: AWWA, Denver.

AWWA, 2016b. M36—*Water Audits and Loss Control Programs*, 4th ed. Denver, Colo.: AWWA, Denver.

AWWA, 2016c. M44—*Distribution Valves: Selection, Installation, Field Testing, and Maintenance*, 3rd ed. Denver, Colo.: AWWA, Denver.

AWWA, 2016d. M51—*Air Valves: Air-Release, Air/Vacuum, and Combination*, 2nd ed. Denver, Colo.: AWWA, Denver.

AWWA, 2016e. *Benchmarking Performance Indicators for Water and Wastewater Utilities: 2016 Edition*. Denver, Colo.: AWWA, Denver. ISBN 9781625761972.

AWWA, 2016f. *2016 AWWA Compensation Survey: Large Water and Wastewater Utilities*. Denver, Colo.: AWWA, Denver. ISBN 9781625761552.

AWWA, 2016g. *2016 AWWA Compensation Survey: Medium-Sized Water & Wastewater Utilities*. Denver, Colo.: AWWA, Denver. ISBN 9781625761569.

AWWA, 2016h. *2016 AWWA Compensation Survey: Rural Water & Wastewater Utilities*. Denver, Colo.: AWWA, Denver. ISBN 9781625761576.

AWWA, 2016i. *2016 Water and Wastewater Rate Survey - Book*. Denver, Colo.: AWWA, Denver. ISBN 9781625762078.

AWWA, 2015. M14—*Backflow Prevention and Cross-Connection Control: Recommended Practices*, 4th ed. Denver, Colo.: AWWA, Denver.

AWWA, 2014a. M22—*Sizing Water Service Lines and Meters*, 3rd ed. Denver, Colo.: AWWA, Denver.

AWWA, 2014b. M28—*Rehabilitation of Water Mains*, 3rd ed. Denver, Colo.: AWWA, Denver.

AWWA, 2013a. M27—*External Corrosion Control for Infrastructure Sustainability*, 3rd ed. Denver, Colo.: AWWA, Denver.

AWWA, 2013b. M42—*Steel Water-Storage Tanks*, revised ed. Denver, Colo.: AWWA, Denver.

AWWA, 2013c. M56—*Nitrification Prevention and Control in Drinking Water*, 2nd ed. Denver, Colo.: AWWA, Denver.

AWWA, 2012a. *M6—Water Meters—Selection, Installation, Testing, and Maintenance,* 5th ed. Denver, Colo.: AWWA, Denver.

AWWA, 2012b. *M32—Computer Modeling of Water Distribution Systems,* 3rd ed. Denver, Colo.: AWWA, Denver.

AWWA, 2012c. *Standard Methods for Examination of Water and Wastewater.* 22nd ed. (E.W. Rice, R.B. Baird, A.D. Eaton, & L.S. Clesceri, editors). Denver, Colo.: AWWA, Denver. ISBN 978-0-87553-013-0.

AWWA, 2011. *M58—Internal Corrosion Control in Water Distribution Systems.* Denver, Colo.: AWWA, Denver.

AWWA, 2007. Partnership for Safe Water Guidelines for Phase IV Application for the "Excellence in Water Treatment" Award February 2003 (Revised December 2007).

AWWA, 2006a. *M20—Water Chlorination and Chloramination Practices and Principles,* 2nd ed. Denver, Colo.: AWWA, Denver.

AWWA, 2006b. *M33—Flowmeters in Water Supply,* 2nd ed. Denver, Colo.: AWWA, Denver.

AWWA, 2001a. *M2—Instrumentation and Control Manual,* 3rd ed. Denver, Colo.: AWWA, Denver.

AWWA, 2001b. *M19—Emergency Planning for Water Utilities,* 4th ed. Denver, Colo.: AWWA, Denver.

AWWA, 2005. *Water Distribution System Assessment Workbook* (Smith, C. editor). Denver, Colo.: AWWA, Denver.

Bayless, W. & Andrews, R.C., 2008. Biodegradation of Six Haloacetic Acids in Drinking Water. *Journal of Water and Health,* 6:1:15.

Benjamin, M.M., 2014. *Water Chemistry,* 2nd ed. Long Grove, Ill.: Waveland Press Inc., Long Grove, Ill.

Besner, M.C.; Gauthier, V.; Trepanier, M.; Martel, K.; & Prevost, M., 2007. Assessing the Effect of Distribution System O&M on Water Quality. *Journal AWWA,* 99:11:77.

Black & Veatch Corporation & White, G.C., 2010. *White's Handbook of Chlorination and Alternative Disinfectants,* 5th ed. ISBN 9780470180983.

Boyd, G.R.; Dewis, K.M.; Korshin, G.V.; Reiber, S.H.; Schock, M.R.; Sandvig, A.M.; & Giani, R., 2008. Effects of Changing Disinfectants on Lead and Copper Release. *Journal AWWA,* 100:11:75.

Brandt, M.; Clement, J.; & Powell, J., 2004. AwwaRF #91006F. Managing Distribution Retention Time to Improve Water Quality, Phase I & Phase II. Denver, Colo.: AWWA Research Foundation.

Burlingame, G.A., 2007. Ratcheting Up Lab Response to Water Quality Warnings. *Opflow,* 33:2:24.

Burton, F., 1996. *Water and Wastewater Industries Characteristics and Energy Management Opportunities.* Report CR-106941. Palo Alto, Calif.: Electric Power Research Institute. Calif

Carlson, S. & Walburger, A., 2007. AwwaRF #91201. Energy Index Development for Benchmarking Water and Wastewater Utilities. Denver, Colo.: AWWA Research Foundation.

CFR Title 40—Chapter 1, Subchapter D.

Chadderton, R.A.; Christensen, G.L.; Henry-Unrath, P., 1992. AWWAwwaRF #515. Implementation and Optimization of Distribution Flushing Programs. Denver, Colo.: AWWA Research Foundation.

Chang, Y.; Reardon, D.J.; Kwan, P.; Boyd, G.; Brant, J.; Rakness, K.L.; & Furukawa, D., 2008. WRF #3056. Evaluation of Dynamic Energy Consumption of Advanced Water and Wastewater Treatment Systems. Denver, Colo.: Water Research Foundation.

Chapman, D.N.; Ng, P.C.F.; & Karri, R., 2007. Research Needs for On-line Pipeline Replacement Techniques. *Tunnelling and Underground Space Technology,* 22:5--6:503.

Conlon, T.; Weisbrod, G.; & Samiullah, S., 1999. We've Been Testing Water Pumps for Years—Has Their Efficiency Changed? Proc. 1999 ACEEE Summer Study on Energy Efficiency in Industry.

Deb, A.K.; Grablutz, F.; Hasit, Y.; Snyder, J.; Loganathan, G.; & Agbenowsi, N., 2002. AwwaRF #459. Prioritizing Water Main Replacement and Rehabilitation. Denver, Colo.: AWWA Research Foundation.

Deb, A.K.; Momberger, K.A.; Hasit, Y.J.; & Grablutz, F.M., 2000. *Guidance Management of Distribution System Operations and Maintenance.* Denver, Colo.: AWWA and AwwaRF.

Deb, A.K.; Hasit, Y.J.; Grablutz, F.M., & Weston Inc. 1995. AwwaRF #804. Distribution System Performance Evaluation.

Dufresne, L., 2016. Energy Management for Water Utilities. Denver, Colo.: AWWA, Denver.

Easton Consultants, 1995. Strategies to Promote Energy-Efficiency Motors Systems in North America's OEM Markets. Easton Consultants: Stamford, Conn.

Edwards, M. & Triantafyllidou, S., 2007. Chloride-to-Sulfate Mass Ratio and Lead Leaching to Water. *Journal AWWA,* 99:7:96.

EPRI (Electric Power Research Institute Report), 2002. *Water and Sustainability,* volume 4. US Electricity Consumption for Water Supply and Treatment—The Next Half Century. 1006787. Palo Alto, Calif.: EPRI.

European Commission, 2001. Study on Improving the Energy Efficiency of Pumps. AEAT-6559/v5.1.

Fanner, M.; Thornton, J.; Liemberger, R.; & Sturm, R., 2007a. AwwaRF #91163. Evaluating Water Loss and Planning Loss Reduction Strategies. Denver, Colo.: AWWA Research Foundation.

Fanner, M.; Thornton, J.; Liemberger, R.; & Sturm, R.; Davis, S.; & Hoogerwerf, T., 2007b. AwwaRF #91180. Leakage Management Technologies. Denver, Colo.: AWWA Research Foundation.

Fleming, K.K.; Gullick, R.W.; Dugandzic, J. P. ; & LeChevallier, M.W., 2006. AwwaRF #3008. Susceptibility of Distribution Systems to Negative Pressure Transients. Denver, Colo.: AWWA Research Foundation.

Foundation for Cross-Connection Control and Hydraulic Research, 2012. *Manual of Cross-Connection Control,* 10th edition. Los Angeles, Calif.: University of Southern California,. Los Angeles, Calif.

Friedman, M.J.; Hill, A.; Korshin, G.; Valentine, R.; & Reiber, S., 2010. WRF #3118. Assessment of Inorganics Accumulation in Drinking Water System Scales and Sediments. Denver, Colo.: Water Research Foundation.

Friedman, M.; Kirmeyer, G.; Lemieux, J.; LeChevallier, M.; Seidl, S.; & Routt, J., 2010. WRF #4109. Criteria for Optimized Distribution Systems. Denver, Colo.: Water Research Foundation.

Friedman, M.; Kirmeyer, G.; Pierson, G.; Harrison, S.; Martel, K.; Sandvig, A.; & Hanson, A., 2005. AwwaRF #2875. Development of Distribution System Water Quality Optimization Plans. Denver, Colo.: AWWA Research Foundation.

Friedman, M.,; L. Radder, L.; S. Harrison, S.; D. Howie, D.; M. Britton, M.; G. Boyd, G.; H. Wang, H.; R. Gullick, R.; LeChevallier, M.;M. LeChevallier, D. Wood, D.; &and J. Funk, J., 2004. AwwaRF #2686. Verification and Control of Low Pressure Transients in Distribution Systems. Denver, Colo.: AWWA Research Foundation.

Friedman M.; Martel, K.; Hill, A,.; Holt, D.; Smith, S.; Ta, T.; Sherwin, C.; Hiltebrand, D.; Pommerenk, P.; Hinedi, Z.; & Camper, A., 2003. AwwaRF #2606. Establishing Site-Specific Flushing Velocities. Denver, Colo.: AWWA Research Foundation.

GLUMRB (Great Lakes Upper Mississippi River Board of State and Provincial Public Health and Environmental Managers), 2007. Recommended Standards for Water Works. Albany, N.Y.: Health Research, Inc., Albany, N.Y.

Grayman, W.M.; Rossman, L.A.; Arnold, C.; Deininger, R.A.; Smith, C.; Smith, J.F.; & Schnipke, R., 2000. AwwaRF #260. Water Quality Modeling of Distribution System Storage Facilities. Denver, Colo.: AWWA Research Foundation.

Grayman, W.M.; Rossman, L.A.; Deininger, R.A.; Smith, C.D.; Arnold, C.N.; & Smith, J.F., 2004. Mixing and Aging of Water in Distribution System Storage Facilities. *Journal AWWA,* 96:9:70.

Grigg, N.S., 2004. AwwaRF #91025F. Assessment and Renewal of Water Distribution Systems. Denver, Colo.: AWWA Research Foundation.

Gullick, R.W.; LeChevallier, M.W.; Case, J.; Wood, D.J.; Funk, J.E.; & Friedman, M.J., 2005. Application of Pressure Monitoring and Modeling to Detect and Minimize Low Pressure Events in Distribution Systems. *Journal of Water Supply Research & Technology–Aqua.* 54:2:65.

Hall, J.; Zaffiro, A.D.; Marx, R.B.; Kefauver, P.C.; Krishnan, E.R.; Haught, R.C.; & Herrmann, J.G., 2007. On-line Water Quality Parameters as Indicators of Distribution System Contamination. *Journal AWWA*, 99:1 (66-77).

Hasit, Y.J.; DeNadai, A.J.; Gorrill, H.M.; McCammon, S.B.; Raucher, R.S.; & Witcomb, J., 1999. AwwaRF #2605. Cost and Benefit Analysis of Flushing. Denver, Colo.: AWWA Research Foundation.

Karassik, I.J.; Krutzsch, W.C.; Fraser, W.H.; & Messina, J.P., editors, 1976. *Pump Handbook.* New York, N.Y.: McGraw-Hill Book Co.

Kirmeyer, G.J.; Thomure, T.M.; Rahman, R.; Marie, J.L.; LeChevallier, M.W.; Yang, J.; Hughes, D.M.; & Schneider, O., 2014. WRF #4307. Effective Microbial Control Strategies for Main Breaks and Depressurization. Denver, Colo.: Water Research Foundation.

Kirmeyer, G.J.; Friedman, M.; Martel, K.; Thompson, G.; Sandvig, A.; Clement, J.; & Frey, M., 2002. AwwaRF #90882. Guidance Manual for Monitoring Distribution System Water Quality. Denver, Colo.: AWWA Research Foundation.

Kirmeyer, G.J.; Friedman, M.; Martel, K.; Howie, D.; LeChevallier, M.; Abbaszadegan, M.; Karim, M.; Funk, J.; & Harbour, J., 2001. AwwaRF #436. Pathogen Intrusion into the Distribution System. Denver, Colo.: AWWA Research Foundation.

Kirmeyer, G.J., Friedman, M.; Clement, J.; Sandvig, A.; Noran, P.; Martel, K.; Smith, D.; LeChevallier, M.; Volk, C.; Antoun, E.; Hiltebrand, D.; Dyksen, J.; & Cushing, R., 2000. AwwaRF #90798. Guidance Manual for Maintaining Distribution System Water Quality. Denver, Colo.: AWWA Research Foundation.

Kirmeyer, G.J.; Kirby, L.; Murphy, B.M.; Noran, P.F.; Martel, K.D.; Lund, T.W.; Anderson, J.L.; & Medhurst, R., 1999. AwwaRF #254. Maintaining Water Quality in Finished Water Storage Facilities. Denver, Colo.: AWWA Research Foundation.

Klein, G., 2005. *California's Water Energy Relationship.* Final Staff Report, CEC-700-2005-011-SF. Sacramento, Calif.: California Energy Commission.

Larock, B.E.; Jeppson, R.W.; & Watters, G.Z., 2000. *Hydraulics of Pipeline Systems.* New York, N.Y.: CRC Press,. New York

LeChevallier, M.W., 2014. WRF #4321. Pressure Management: Industry Practices and Monitoring Procedures. Denver, Colo.: Water Research Foundation.

LeChevallier, M.W.; Gullick, R.W.; Karim, M.R.; Friedman, M.J.; & Funk, J.E., 2003. The Potential for Health Risks From Intrusion of Contaminants Into the Distribution System From Pressure Transients. *Journal of Water Health,* 1:1:3.

Lee, J.J.; Schwartz, P.; Sylvester, P.; Crane, L.; Haw, J.; Chang, H.; & Kwon, H.J., 2003. AwwaRF #90928F. Impacts of Cross-Connections in North American Water Supplies. Denver, Colo.: AWWA Research Foundation.

Linder, K. & Martin, B., 2015. *Self-Assessment for Water Treatment Plant Optimization.* Denver, Colo.: AWWA, Denver.

Lytle, D.A., 2007. Accumulation of Contaminants in the Distribution Systems. *USEPA ORD/OGWDW Workshop on Inorganic Contaminant Issues.* Cincinnati, Ohio: USEPA.

Lytle, D.A. & Schock, M.R., 2008. The Relationship Between Redox Stability, and Corrosion and Metal Release. Proc. AWWA 2008 WQTC Conf., Cincinnati, Ohio.

Lytle, D.A. & Schock, M.R., 2005. Formation of Pb (IV) Oxides in Chlorinated Water. *Journal AWWA,* 97:11:102.

Mahmood, F.; Pimblett, J.; Hill, C.; & Chowdhury, Z., 2009. CFD Modeling and Spreadsheet Tool to Evaluate Mixing and Water Quality in Storage Tanks. Proc. AWWA 2009 Annual Conference and Exhibits, San Diego, Calif.

Mahmood, F.; Pimblett, J.G.; Grace, N.O.; & Grayman, W.M., 2005. Evaluation of Water Mixing Characteristics in Distribution System Storage Tanks. *Journal AWWA,* 97:3:74.

Maier, P.; White, R.; Connell, S.; Kroll, K.; King, G.; Hanley, G.; & Metzger, R., 2008. Energy Savings Through Pump Refurbishment and Coating. *Pumps & Systems.*

Makar, J.; Rogge, R.; McDonald, S.; & Tesfamariam, S., 2005. AwwaRF #91053. The Effect of Corrosion Pitting on Circumferential Failures in Grey Cast Iron Pipes. Denver, Colo.: AWWA Research Foundation.

NASTB, 2006. Drinking Water Distribution Systems: Assessing and Reducing Risks. Washington D.C.: National Research Council Report.

Nguyen, C.K.; Stone, K.R.; & Edwards, M.A., 2011. Chloride-to-Sulfate Mass Ratio: Practical Studies in Galvanic Corrosion of Lead Solder. *Journal AWWA,* 103:1:81.

NRC (National Research Council), 2006. *Drinking Water Distribution Systems: Assessing and Reducing Risks.* Washington D.C.: National Academies Press,. Washington D.C.

Pereira, V.; Weinberg, H.S.; & Singer, P.C., 2004. Temporal and Spatial Variability of DBPs in a Chloraminated Distribution System. *Journal AWWA,* 96:11:91.

Renner, R.C. & Hegg, B., 1997. AwwaRF #274. Self-Assessment Guide for Surface Water Treatment Plant Optimization. Denver, Colo.: AWWA Research Foundation.

Rishel, J.B., 2002. *Water Pumps and Pumping Systems.* New York, N.Y.: McGraw-Hill Professional., New York

Roberts, P.J.W.; Tian, X.; Sotiropoulos, F.; & Duer, M., 2006. AwwaRF #2898. Physical Modeling of Mixing in Water Storage Tanks. Denver, Colo.: AWWA Research Foundation.

Rossman, L.A.; Clark, R.M.; & Grayman, W.M. 1994. Modeling Chlorine Residuals in Drinking-Water Distribution Systems. *Journal of Environmental Engineering,* 120:4:803.

Routt, J.C.; Sekhar, M.; & Friedman, M., 2009. Optimizing Drinking Water Distribution Systems for Disinfection and Disinfection Byproducts. Proc. WEF/AWWA 2009 Disinfection Conf., Atlanta, Ga.

Schneider, O.D.; Bukhari, Z.; Hughes, D.; Fleming, K.; LeChevallier, M.W.; Schwartz, P.; Sylvester, P.; & Lee, J.J., 2009. WRF #3022. Cross-Connection and Backflow Vulnerability: Monitoring and Detection. Denver, Colo.: Water Research Foundation.

Schock, M.R.; Hyland, R.; & Welch, M. 2008. Occurrence of Contaminant Accumulation in Lead Pipe Scales fFrom Domestic Drinking Water Distribution Systems. *Environmental Science and Technology,* 42:12:4285.

Schock, M.R., 2005. Distribution Systems and Reservoirs and Reactors for Inorganic Contaminants. *Distribution System Water Quality Challenges in the 21st Century.* Denver, Colo.: AWWA., Denver.

Singley, J.E., 1981. The Search for a Corrosion Index. *Journal AWWA,* 73:11:578.

States, S., 2010. *Security and Emergency Planning for Water and Wastewater Utilities.* Denver, Colo.: AWWA, Denver.

Sturm, R.; Gasner, K.; Wilson, T.; Preston, S.; & Dickinson, M.A., 2014. WRF #4372a Real Loss Component Analysis: A Tool for Economic Water Loss Control.

Thorley, A.R.D., 2004. *Fluid Transients in Pipeline Systems,* 2nd ed. D. & L. George, Ltd., Herts, England.

USDOE (US Department of Energy). Energy Efficiency and Renewable Energy. Best Practices. Pumping Tip Sheets.

USEPA. 2017. Effective Utility Management: A Primer for Water and Wastewater Utilities. http://www.nacwa.org/docs/default-source/resources---public/eum-primer-final-1-24-17.pdf

USEPA (US Environmental Protection Agency). CUPSS: Check-Up Program for Small Systems.

USEPA, 2016. Optimal Corrosion Control Treatment Evaluation Technical Recommendations for Primacy Agencies and Public Water Systems. EPA 816-B-16-003.

USEPA, 2008a. Total Coliform Rule/Distribution System Federal Advisory Committee (TCRDS FAC). Agreement in Principle. Agreement in Principle.

USEPA, 2008b. National Secondary Drinking Water Regulations CFR 40 Part 143. e-CFR.

USEPA, 2008c. EPANET.

USEPA, 2008d. Water Security Initiative: Interim Guidance on Developing an Operational Strategy for Contamination Warning Systems.

USEPA, 2008e. Water Security Initiative: Interim Guidance on Developing Consequence Management Plans for Drinking Water Utilities.

USEPA, 2008f. Water Quality in Small Community Distribution Systems Reference Guide for Operators. EPA/600/R-08/039CD.

USEPA, 2008g. Asset Management: A Best Practices Guide. EPA 816-F-08-014.

USEPA, 2007. Simultaneous Compliance Guidance Manual for the Long Term 2 and Stage 2 DBP Rules. EPA 815-R-07-017.

USEPA, 2006a. Distribution System Indicators of Water Quality. Total Coliform Rule Issue Paper

USEPA, 2006b. The Effectiveness of Disinfectant Residuals in the Distribution System. Total Coliform Rule Issue Paper.

USEPA, 2006c. Initial Distribution System Evaluation Guidance Manual for the Final Stage 2 Disinfectants and Disinfection Byproducts Rule. EPA 815-B-06-002. Washington, D.C.

USEPA, 2006d. Stage 2 Disinfectants and Disinfection Byproducts: Final Rule. *Federal Register,* 71:2:388.

USEPA, 2006e. Inorganic Contaminant Accumulation in Potable Water Distribution Systems. Total Coliform Rule Issue Paper. Washington, D.C.

USEPA, 2006f. Interactive Sampling Guide for Drinking Water System Operators EPA816-F-03-016.

USEPA, 2005. WaterSentinel Online Water Quality Monitoring as an Indicator of Drinking Water Contamination. Draft, version 1.0 EPA 817-D-05-002, EPA 817-D-05-001.

USEPA, 2004. Response Protocol Toolbox and Guidelines: Planning for and Responding to Drinking Water Contamination Threats and Incidents Module 1: Water Utilities Planning Guide. Interim final August 2004. Office of Ground Water and Drinking Water, Water Security Division EPA 817-D-04-001 2003, EPA 817-D-04-001 2004.

USEPA, 2003. Revised Guidance Manual for Selecting Lead and Copper Control Strategies EPA-816-R-03-001.

USEPA, 2002a. Long Term 1 Enhanced Surface Water Treatment Rule. *Federal Register,* 67:9:1811.

USEPA, 2002b. Potential Contamination Due to Cross-Connections and Backflow and Associated Health Risks EPA White Paper, Washington D.C.

USEPA, 2001. Controlling Disinfection By-Products and Microbial Contaminants in Drinking Water. EPA/600/R-01/110.

USEPA, 1998a. Interim Enhanced Surface Water Treatment Rule. *Federal Register,* 63:241:69478.

USEPA, 1998b. Bender, Jon H., Bissonette, Eric M., DeMers, Larry D., Hegg, Bob A., and Lieberman, Richard J. 1998. *Handbook for Optimization of Water Treatment Plants Using the Composite Correction Program.* EPA/625/6-91/027

USEPA, 1998c. Stage 1 Disinfectants and Disinfection Byproducts: Final Rule. *Federal Register,* 63:2421:69390.

USEPA, 1988a, 1991, 2004. Lead and Copper Rule & Guidance.

USEPA, 1988b. Lead Copper Rule. Final Rule. *Federal Register,* 72:195:57782.

USEPA, 1989a. Surface Water Treatment Rule. *Federal Register,* 54:24:27486.

USEPA, 1989b. Total Coliform Rule. Final Rule. *Federal Register,* 54:124:27544.

Vasconcelos, J.; Boulos, P.; Grayman, W.; Kiene, L.; Wable, O.; Biswas, P.; Bhari, A.; Rossman, L.; Walski, T.M.; Chase, D.V.; Savic, D.A.; Grayman, W.; Beckwith, S.; & Koelle, E., 2003. *Advanced Water Distribution Modeling and Management.* Waterbury, Conn.: Haestad Press, Waterbury, Conn.

Vasconcelos, J.; Boulos, P.; Grayman, W.; Kiene, L.; Wable, O.; Biswas, P.; Bhari, A.; Rossman, L.; Clark, R.; & Goodrich, R., 1996. Characterization and Modeling of Chlorine Decay in Distribution Systems. Denver, Colo.: AWWA and AwwaRF.

Verosky, K.; Maier, P.; White, R.; Connell, S.; Kroll, K.; King, C.; Hanley, G.; & Metzger, R., 2008. *Energy Savings Through Pump Refurbishment and Coating.* Pumps & Systems.

Walski, T.M.; Chase, D.V.; Savic, D.A.; Grayman, W.; Beckwith, S.; & Koelle, E., 2003. *Advanced Water Distribution Modeling and Management.* Waterbury, Conn.: Haestad Press, Waterbury, Conn.

Whelton, A.J.; Dietrich, A.M.; Gallagher, D.L.; & Roberson, J.A., 2007. See my response.

Using Customer Feedback for Improved Water Quality and Infrastructure Monitoring. *Journal AWWA,* 99:11:62.

White, G.C., 1999. *Handbook of Chlorination and Alternative Disinfectants,* 4th ed. Danvers, Mass.: John Wiley & Sons, Inc., Danvers, Mass. ISBN 0-471-29207-9.

Zhang, Y.; Love, N.; & Edwards, M., 2009. Nitrification in Drinking Water Systems. *Critical Reviews in Environmental Science Aand Technology,* 39:3:153.

Zhang, X. & Minear, R.A., 2002. Decomposition of Trihaloacetic Acids and Formation of the Corresponding Trihalomethanes in Drinking Water. *Water Research,* 36:14:3665.

SPECIAL STUDY EXAMPLE

Special studies are designed to help distribution system staff take steps to more closely evaluate a specific distribution system issue or perform the activities and data collection needed to evaluate the need for additional operational modifications. They may also be used to evaluate and optimize processes or operations, document past performance, modify system control activities, justify administrative procedures or policy changes, or to identify design changes necessary to improve performance.

Special studies generally consist of the following steps:

- Problem definition
- Hypothesis
- Approach, including steps taken and data collection requirements
- Duration
- Expected results, including how to evaluate data and how success is defined
- Conclusions
- Implementation

They provide a structured, systematic approach of evaluating system operating conditions following scientific method. Documenting a special study consists of collecting the necessary data and producing a written summary that defines the project purpose, approach, duration of the study, expected results, conclusions, and implementation plan. An example special study format is displayed in Table A-1. If a special study were applied during the self-assessment process, utilities should consider including the details of the study, including the outcome and application, in the self-assessment completion report.

As described in Table A-1, the purpose for or the expected outcome (hypothesis) of the Special Study should be sufficiently narrow in scope and should clearly define the study to be conducted. The description of the approach and duration should include information including when and where samples are to

Table A-1. Special study format

Special Study
Special Study Name:
Project Purpose: Describe the purpose of the project, highlight the cause and effect relationship that is anticipated. Be sure to keep the definition of the project purpose sufficiently narrow in scope such that data can be readily collected and success can be readily defined.
Approach: Provide a detailed procedure describing how the study will be conducted. Include information about data collection, operational changes that may be required, and responsible parties. Ensure that staff are involved in the development of the Special Study procedure, particularly if they will be directly involved with the study, which will help to enhance staff buy-in of the study's procedures and application.
Duration of Study: Define the study's duration and the time after which results will be evaluated.
Expected Results: Describe the projected results of the study, which may help to better define the success or limitations of the Special Study effort. At times, the outcomes of the Special Study may be unknown at the time of the study's initiation. This is acceptable, and the expected outcome may be an estimate or projection of the anticipated impact. The expected results may not be consistent with the actual results. This is also acceptable and does not constitute a failure of the Special Study process.
Conclusions: The study's conclusions and impact should be documented (including data), so that the results of the Special Study effort may be used as a training tool for appropriate utility staff. Documenting the study's conclusions also provides a record of the approach that was used to address the issue and the outcomes of this approach, which provides a quantifiable record for future work.
Implementation: Describe the outcome of the Special Study in terms of changes implemented as a result. Be sure to describe and/or justify changes made to operational procedures that are supported by the results of the study. This helps to formalize the mechanisms used to improve distribution system performance.

be collected, who is to collect the samples, what analyses are to be conducted and over what period, and how the results are to be analyzed and presented. This approach should be developed in cooperation with the operational staff to obtain commitment and to develop a high-quality procedure that can be followed consistently by distribution system staff.

It is important that the study results be documented by compiling data into graphs, figures, and/or tables for interpretation. The results of every Special Study may not support the original hypothesis and/or expected outcome. This is a very real part of science—sometimes unexpected results actually help to facilitate more learning than results are routine and anticipated.

Documenting results, as described previously, allows the findings to be presented to staff, administrators, regulatory agencies, or other stakeholders as a basis for implementing changes in system operation, design, maintenance, or administration leading to improved performance. An implementation plan, together with documentation, addresses procedural changes and support required to implement modifications that have developed as an outcome of the special study results. The implementation plan should be brief and generalized until the results of the study show that the suggested solution is a viable option. If all of these steps are followed, the Special Study approach ensures involvement by the operational staff, serves as a basis for ongoing training and increases confidence in operator capabilities.

ACTION PLANNING

As described in chapter 7, after the identification and prioritization of performance limiting factors, action plans should be developed for the highest priority performance limiting factors, to provide the utility with a structured framework and plan for working toward optimized operations and performance. The Topic Development Sheet, displayed in Table A-2, provides a series of questions that can help the self-assessment team structure a discussion around a particular performance limiting factor in order to develop specific actions. Note that this example is provided as a resource only and is not a required component of the Phase III Self-Assessment Completion Report.

Table B-1. Example topic development sheet

TOPIC DEVELOPMENT SHEET	
Topic/Issue:	
Benefits of Addressing this Issue:	
Potential Obstacles:	Potential Solutions:
Potential Action Items (short and long term):	

The most appropriate potential actions identified in the discussion and summarized in the table prepared by the utility are transferred to the utility's action plan. Creation of the action plan is one of the primary outcomes of the self-assessment process. The action plan lists both short- and long-term items to be completed, including the responsible party and projected due date. The action plan is a living document that should be reviewed and updated on a regular basis to help ensure progressive implementation of performance improvement activities. Partnership for Safe Water subscriber utilities that have attained Phase III status provide an optimization update as a component of the program's annual reporting process. Table A-3 provides an example of an action plan table. Examples of completed action plans for various components of the distribution system self-assessment are provided in chapter 7. Utilities may use this action plan table format to create action plans for the Phase III Self-Assessment Completion Report or for internal purposes.

Table B-2. Example action plan table

Issue	Short-term solution	Person(s) responsible	Target date to be completed	Long-term solution	Person(s) responsible	Target date to be completed

PARTNERSHIP FOR SAFE WATER: PHASE III SELF-ASSESSMENT COMPLETION REPORT EVALUATION PARAMETER DEFINITIONS

The following parameters are used by the Partnership's PEAC (peer-review team) in the evaluation of the Phase III Self-Assessment Completion Report. Parameters are grouped under the primary categories of Performance, Administrative, Operational, Design, and Overall. Definition for the evaluation parameters are provided in this Appendix. Although some of these parameters may relate to specific sections of the self-assessment report, information for any section of the report may be used in the evaluation of a specific scoring parameter. These definitions are provided to assist utilities preparing a self-assessment completion report with a better understanding of how the report is evaluated during the peer-review process. Additional information about the self-assessment and peer-review process may be obtained by contacting the Partnership for Safe Water.

Performance Parameters:

- *Accurately Assessed Performance Compared withto Partnership Optimization Goals*—The participant has accurately documented their current status compared with Partnership optimization goals. Meeting the performance goals in Phase III is not required; however, striving to meet all established Partnership goals (e.g., disinfectant residual, pressure, main break frequency) via routine review of adequate and accurate data is a very critical component because it serves as the primary basis for action plan development. The system must provide results of its data interpretation, document that they are based on valid measurements, verify that existing data reflects the actual status of the distribution system, and consider where additional data are needed. Reviewer considerations should include completeness of data submittals for all major assessment areas, consistency between results and data included in assessment spreadsheets, and an assessment to determine if the data are representative and adequate. It is expected that the self-assessment process will identify the need to obtain more representative data at additional and specific locations throughout the distribution system.

- *Provided Performance Benchmark Values (for required factors)*—Several performance-related benchmark values (e.g., disinfection, pressure) are required, and these are described in the Self-Assessment Guide and included in the Optimization Assessment Tool. Reporting of specific financial benchmarks is not required because of the sensitive nature of this information. If financial benchmarks are not provided, the utility should provide substitute data and information that supports the long-term financial viability of the utility. Reviewers should examine these values for completeness and accuracy. *The performance-limiting factor priority ranking and action plan development should address all benchmark values that are not optimal as well as highlight utility actions toward achieving long-term optimization and financial viability.*

- *Understand/Explain Performance Deviations*—The participant has demonstrated the capability to discuss deviations from optimum performance goals in a logical and convincing manner. The submittal should demonstrate an understanding of deviations and should impart an urgency to rectify deviations, both short- and long-term. Higher ratings should be considered when deviations are identified in a timely manner through routine review of performance data trends by key staff at multiple levels of the organization. A lower rating should be given if no mention is made to address performance deviations in the data submitted or higher priority is not indicated where deviations from optimum are indicated.

Administrative Parameters:

- *Acceptance of Optimization Goals by Staff at All Levels*—The participant should demonstrate a commitment to achieving and recognition of the value of the optimization goals. The submittal should provide examples of how recognition of the public health benefits and acceptance of pursuing a goal "beyond the regulations" has become a priority for staff at all levels of the utility, including distribution system operations and maintenance staff. The report content should document that optimization goal recognition and acceptance is apparent throughout the organization in daily decision-making, and has been incorporated into overall mission statement. Systems may establish their own performance goals in areas not covered by the Partnership or as interim goals as they continue to progress toward Partnership goals. There should be a clear indication that all system personnel are actively striving to meet interim goals now, and ultimately Partnership goals in the future. Higher ratings should be considered for reports that clearly convey examples of how all key optimization goals are considered by distribution system operations and maintenance staff as they go about their duties.

- *Involvement of All Parties in the Partnership Process*—Key personnel at all levels of administration, Operation, and Maintenance must be involved in the Partnership effort. The role of Administrators in setting goals and providing an environment that expects and encourages involvement in optimization at all levels of distribution system operation and maintenance is essential for success. The submittal should indicate this involvement in addressing all aspects of the Self-Assessment Procedures. The participant should also demonstrate that a multilevel involvement occurred during specific optimization activities and during development of the Completion Report submittal. Understanding of the optimization goals and optimization process is necessary to ensure buy-in at *every level*, including the distribution system operations and maintenance staff responsible for daily water quality monitoring, repair, and replacement of infrastructure. Reviewers should examine the self-assessment team roster to evaluate the degree of involvement of all parties. A higher rating should be considered for submittals that include information from routine optimization meetings and special studies documenting that multiple staff (e.g., administration, operation, maintenance) were actively involved in the self-assessment process.

- *Documentation/Demonstration of Addressing Complacency*—The tenacity to not only achieve optimization goals but to maintain the optimized level of performance under real or potential variations (e.g., main breaks) should be indicated. A newer system that provides relief from some of the issues facing older systems should include recognition of these issues and provide for them as system components age. The submittal should describe approaches to maintain optimal water quality while addressing nonroutine, unexpected, and/or "emergency situations" whether they are intentional, natural disasters, or construction mishaps. The ability to rally a technically sound response to protect the system integrity under difficult situations, especially emergency repair of main breaks, should be demonstrated. Higher ratings should be considered for submittals which include system-specific standard operating procedures (SOPs) that are well written and routinely updated to address emergency and nonroutine situations with consideration for maintaining optimal water quantity and quality. Overall, the submittal should convey that proactive decision-making is evident throughout the organization.

- *Commitment to Staffing/Funding Resources*—The participant should demonstrate that resources of staffing and funding are available or are being pursued to support the optimization activities. If limitations are identified, an approach and a realistic time schedule to address resource limitations should be described. The participant should also demonstrate a commitment to support optimization activities even though major changes may be planned for the future. Reviewers should assess the administrative benchmarks for evidence that adequate support is being provided. Nonoptimized indications for these performance limiting factors should result in higher priority ranking.

Operational Parameters:

- *Documented Application of Operational Control Skills*—The participant should demonstrate, through ranking of status questions with nonoptimized indications, the operational and management procedures used to continuously provide optimal water quality to all portions of the distribution system. The report should clearly outline how all distribution system operations and maintenance staff routinely implement these procedures; and that procedures are updated as necessary. Data collection, interpretation, and associated adjustments should be indicated by established goals and performance results and action plans to improve to meet the goals. Judgment and experience of the reviewers will be especially important in this area. A "reading between the lines" will be a key evaluation effort for this parameter. This is the parameter in which reviewers should carefully evaluate the required SOPs for corrective flushing, disinfectant residual sampling, and emergency repair of mains. A higher rating may be considered for submittals where these SOPs are well-written, system-specific, routinely updated, and clearly convey consistent application of optimal process control skills relative to operation and management of the distribution system.

- *Priority Setting Capability*—The participant should demonstrate the capability to set priorities on activities that indicate an understanding and commitment to the optimization goals. An explanation of who is responsible for priority setting and the method used to determine priorities should be included and reflect that the ultimate goal of optimal water quality throughout the distribution system is a primary consideration. The inability to develop feasible action plans to remedy high ranking performance limiting factors within a reasonable time frame would indicate a lack of understanding/skill in this area. The ability to pursue design, administration, or maintenance limitations to support optimization, operational, and maintenance needs should be indicated. Special studies of operational limitations included in the action plan indicate an ability to identify and correct problems in a rational and systematic manner. Reviewers may consider a lower score if no special studies are indicated or no actions have been taken to at least begin addressing high-ranking performance-limiting factors where feasible short-term solutions exist.

- *Training/Communications Capability*—The participant should demonstrate the recognition of the need for a formalized system to ensure that adequate communications and understanding are provided throughout the utility. Capability to support the communications and training needs should be demonstrated. Efforts to ensure accurate and consistent responses to necessary process changes should be indicated. Reviewers should examine performance assessment status questions that address training and communications to determine a score for this parameter. Higher scores may be considered if the submittal documents that system-specific/in-house training programs have been developed to address specific needs of staff at their utility.

Design Parameters:

- *Distribution System Modifications Based on Performance Needs*—The participant should demonstrate that distribution system design and physical infrastructure limitations will be addressed in a manner that supports optimization goals. When rating this parameter, reviewers should carefully consider the system's documented asset management program and long-term replacement and renovation plan. Utilization of the Partnership to obtain lower priority distribution system modifications that cannot be tied to performance objectives should result in a lower rating. The expected or realized link between modifications made, or proposed, and performance should be indicated in the documentation if a higher rating is to be assigned; before and after data should be included whenever possible. Higher ratings should be considered for submittals that clearly document data-driven decision-making for established optimization parameters (e.g., disinfectant residual, pressure, main break frequency); more specifically, that key system personnel routinely evaluate system-specific data trends when determining short- and long-term modification needs.

Overall Parameters:

- *Reasonable Prioritization of Performance Limiting Factors*—The report should explain the process the utility used to prioritize limiting factors and the link between prioritization and optimization. The prioritization approach used should be technically sound and should demonstrate an ability to prioritize actions in a manner that will help the system progress toward consistently meeting optimized performance goals. In general, factors currently affecting performance on a routine basis should be considered as highest priority items, factors currently impacting performance on an intermittent basis would be next in the priority ranking, whereas factors not affecting performance now but having potential to impact performance in the future would be ranked lowest. The facility should ultimately use this prioritization process to focus action plan development, as outlined below.

- *Development of Action Plans to Improve Performance*—Action plans are considered to be a primary outcome of the self-assessment process because they outline system-specific plans to address factors limiting optimization. Therefore, all submittals should include feasible action plans for at least the highest ranking performance limiting factors, as identified and outlined in the previously defined rating category (Reasonable Prioritization of Performance Limiting Factors). There is not a specified minimum or maximum number of action items required for the report, as this is a very utility-specific parameter. This parameter will require judgment on the part of the evaluator regarding the feasibility, effectiveness, and timeliness of the proposed action plans. Higher ratings should be considered when detailed action plans designate a variety of individual(s) from various positions as responsible for implementation, include feasible short and long term actions, and include aggressive timelines for completion. The expected outcome of action plans should reflect a link to optimized performance.

- *Convincing/Understandable Description of Changes/Progress*—The reviewer should have a comfortable feeling about the sincerity of the submittal and the magnitude of the changes made or proposed, based on a review of all the documentation. Reviewers should consider that the time needed to implement distribution system improvement plans, and fully realize the impacts of those actions, is significantly greater than when making in-plant process control modifications. Taking this into consideration, reviewers should evaluate this category with regard to progress made on the highest priority performance-limiting factors. It is not adequate for a system to simply report that it is already fully optimized. All submittals should provide convincing documentation for the reviewer to explain the efforts made to reach the current status as well as future plans to maintain optimized water quality and continue to improve.

EXAMPLE RECORD-KEEPING FORM CHLORINE ANALYZER COMPARISON CHECK

As described in chapter 5, it is important to periodically verify and/or calibrate analytical instrumentation that is providing data used for operational control and regulatory reporting. Utility staff should adhere to any regulatory requirements for instrumentation calibration and verification, including the frequency of performing such activities. Be sure to follow the manufacturer's instructions for instrument calibration and verification. The results of instrument calibration and verification activities should be properly recorded and stored for future reference and according to any applicable regulatory requirements.

Because of the importance of disinfectant residual to public health protection, as well as its status as one of the distribution system optimization program's primary optimization parameters, Figure D-1 presents an example Chlorine Analyzer Comparison Check form. This form would be used when a grab sample comparison measurement is used to verify performance of an online chlorine analyzer. Information collected on this form could be collected onsite and then transferred to a utility's electronic data historian or similar electronic system for long-term retention. Although this example presents a form that is specific to chlorine analyzer verification, similar forms could be created for performance verification of instruments that are monitoring for a variety of water quality parameters.

This example record-keeping form for a chlorine analyzer comparison check was provided courtesy of Pennsylvania Department of Environmental Protection.

Method 334.0 Routine Comparison Calibration for Online Chlorine Analyzers

PWSID#: _____   Plant Name: _____   Month/Year: _____

Online Analyzer Make/Model #: _____

Online Analyzer Monitoring Location: _____

Grab Sample Colorimeter Make/Model #: _____

Date of Last Colorimeter QA/QC: _____

Date/ time of comparison check	Initials of operator conducting comparison check	(A) Grab sample result via verified colorimeter (mg/L Cl$_2$)	(B) Online analyzer reading at time of grab sample (mg/L Cl$_2$)	(C) Calculate % difference ((A-B) ÷ A) X100	(D) If Column (C) is greater than (>) 15%, actions taken may be summarized in this column. Potential options include: (1) Conduct Online Analyzer QA/QC troubleshooting & perform another comparison check* (2) If Online Amperometric analyzer, adjust reading to match DPD Grab Sample Result as per instrument manual (3) If Online Colorimetric analyzer, contact the instrument manufacturer for calibration guidance *Contact instrument manufacturer for QA/QC Guidance. Be sure to complete and record second comparison check after any equipment calibration or adjustments

Figure D-1.   Example record-keeping form—Chlorine Analyzer Calibration Check

EXAMPLE STANDARD OPERATING PROCEDURE: DISINFECTANT RESIDUAL SAMPLING AND ANALYSIS

Standard operating procedures (SOPs) are a required component of the distribution system Phase III Self-Assessment Completion Report. The required SOPs are listed in the Phase III Self-Assessment Completion Report Checklist. These required SOPs include disinfectant residual sampling and analysis, emergency repair of mains, and corrective flushing. Utilities may submit additional SOPs as part of the Self-Assessment Completion Report, as appropriate, to support responses to the self-assessment questions.

Provided in this Appendix is an example SOP for disinfectant residual sampling and analysis. The example SOP provided in this Appendix is intended to provide utilities with an example of one type of structure, format, and content that may be used in a set of comprehensive distribution system SOPs. The SOP provided in this Appendix should not be interpreted as a specific recommendation and may not be applicable, as written, to all distribution systems. Distribution system staff is encouraged to develop utility-specific SOPs for internal use. Some best practices relating to SOPs include regularly reviewing and updating SOP documents, training appropriate staff on the procedures included in the SOP, and periodically verifying the effectiveness of the training.

Standard Operating Procedure
Disinfectant Residual Sampling and Analysis
Free Chlorine

Purpose: The purpose of this procedure is to collect a representative distribution system sample for free chlorine and accurately analyze the sample in the field. This procedure is used for routine and compliance sampling from distribution system sampling sites.

Safety:

- Chemicals—read and understand the material safety data sheets for all chemicals involved in testing.

- Field—use the appropriate number of staff when sampling and working in the field. Fieldwork may occasionally involve special hazards that field operators should be aware of and prepared for.

- Laboratory—always follow proper laboratory safety procedures. Use the correct personal protection equipment when working in the laboratory.

Notes:

- Free chlorine is a volatile parameter and must be analyzed onsite to prevent sample degradation.

- Chlorine demand in the containers and glassware used for chlorine analysis can result in false low readings. Ensure that the containers and glassware used for chlorine analysis are clean and rinsed with sample before analysis.

Tools:

- Chlorine portable colorimeter test kit (colorimeter/sample cells)
- Portable colorimeter instrument manual
- Pocket thermometer
- Free chlorine reagents
- Laboratory wipes
- Alcohol wipes
- Free chlorine testing sample form

Procedure:

1. Bring the chlorine test kit to the tap or location that is going to be sampled.

2. If sampling from a faucet, remove aerator from faucet and ensure that faucet is free of strainers, o-rings, hose attachments, or purifying devices.

3. Open an alcohol wipe and wipe the outside of the faucet to disinfect any areas that may have come into contact with contaminants.

4. Open the cold water tap and run a nonsplashing stream of water.

5. Using the thermometer, measure the temperature of the water until it reaches a steady reading. Record the reading on the sample form.

6. Remove sample cell from the colorimeter test kit. Rinse the sample cell and cap three times with the water that is to be tested.

7. From the running tap, fill the sample cell to the 10-mL fill line, ensuring that the bottom of the meniscus is at the fill line on the cell. Put the cap on the sample cell.

8. Wipe the sample cell with a wipe, so that its surface is clean and dry.

9. Place the sample cell into the instrument (be sure to orient the cell consistently for each measurement, using markings on the sample cell as a guide). Fit the meter cover over the sample cell compartment. Zero the instrument using the Zero key.

10. Remove the cell from the instrument and add the free chlorine reagent. Cap the cell, gently mix the sample, and wait for 30 seconds. Wipe the sample cell with a wipe, and insert it into the instrument. Fit the meter cover over the sample cell compartment. Press the Read/Measure key to obtain the free chlorine measurement in mg/L. (Note that the meter should provide a free chlorine reading to two decimal places X.XX mg/L. If only one decimal place is displayed, follow the instructions in the instrument manual, to change the instrument from high range into low range measurement mode.)

11. Record the measurement on the sample form.

12. Empty and rinse the sample cells with tap water, and place the empty sample cells back in the colorimeter kit for their next use.

Goal:

The utility goal for free chlorine measurement is > 0.20 mg/L from all distribution system sample sites.

Response:
In the event that the free chlorine measurement is ≤ 0.20 mg/L at any sample site, notify the plant manager and distribution department staff about the sample site at which the chlorine concentration is low. Treatment adjustments may need to be made or that area of the main may need to be flushed. Record the verbal notification and response taken in the field operator logbook.

SAMPLE CUSTOMER COMPLAINT FORM

An example of the typical information collected to record a customer complaint in included in this Appendix. This is a general example and is not intended to include every parameter that may be recorded in association with customer complaints. Development of a utility-specific means of recording customer complaints and responses is recommended to ensure collection of the most relevant information.

Complaint Date:	Time:	Recorded by:
Customer Name and Account Number:	Address:	Phone:
Nature of Complaint:		
Action Taken:		
Field Employee:		Date:
Time Arrived:		Time Finished:
Sample Taken? Y or N	Bacti:	Chlorine Residual:
Type of Plumbing:		
Size of Water Main:		
Type of Water Main:		
Date Water Main Installed:		
Follow-up Actions Taken (describe or N/A):		
Supervisor Signature:		
Manager Signature:		

INDEX

Note: *f.* indicates figure; *t.* indicates table